Terrorism in Asymmetrical Conflict

Ideological and Structural Aspects

Terrorism in Asymmetrical Conflict

Ideological and Structural Aspects

SIPRI Research Report No. 23

Ekaterina Stepanova

OXFORD UNIVERSITY PRESS
2008

OXFORD

UNIVERSITY PRESS

Great Clarendon Street, Oxford OX2 6DP

Oxford University Press is a department of the University of Oxford.
It furthers the University's objective of excellence in research, scholarship,
and education by publishing worldwide in

Oxford New York

Auckland Cape Town Dar es Salaam Hong Kong Karachi
Kuala Lumpur Madrid Melbourne Mexico City Nairobi
New Delhi Shanghai Taipei Toronto

With offices in

Argentina Austria Brazil Chile Czech Republic France Greece
Guatemala Hungary Italy Japan Poland Portugal Singapore
South Korea Switzerland Thailand Turkey Ukraine Vietnam

Oxford is a registered trade mark of Oxford University Press
in the UK and in certain other countries

Published in the United States
by Oxford University Press Inc., New York

British Library Cataloguing in Publication Data
Data available

Library of Congress Cataloging in Publication Data
Data available

Typeset and originated by Stockholm International Peace Research Institute
Printed and bound in Great Britain on acid-free paper by
Biddles Ltd, King's Lynn, Norfolk

This book is also available in electronic format at
http://books.sipri.org/

ISBN 978–0–19–953355–8
ISBN 978–0–19–953356–5 (*pbk*)

Contents

Preface

Despite the growing scope of terrorism literature, especially since 11 September 2001, some of the toughest questions concerning security threats posed by terrorism remain unanswered. What does asymmetry in conflict mean for terrorism and anti-terrorism efforts? Why is terrorism used as a tactic in some armed conflicts but not others? What are the anti-terrorism implications of dealing with broad armed movements that may selectively resort to terrorist means but, in contrast to some marginal splinter groups, are mass-based and often outmatch in popularity and social activity the weak states where they operate? Why at the same time have relatively small, al-Qaeda-inspired groups challenged and altered the international system so effectively through high-profile terrorism? How is it possible that these small and dispersed cells that are only linked by their shared ideology manage to act as if they were parts of a more structured and coordinated transnational movement?

Breaking new ground, this Research Report provides original insights into these and many other difficult questions. It builds on over a decade of Dr Stepanova's research on terrorism, political violence and armed conflicts. The report looks at the two main ideologies of militant groups that use terrorist means—radical nationalism and religious extremism—and at organizational forms of terrorism at local and global levels, exploring the interrelationship between these ideologies and structures.

Dr Stepanova convincingly concludes that, despite the state's continuing conventional superiority—in terms of power and status—over non-state actors, the critical combination of extremist ideologies and dispersed organizational structures gives terrorist groups many comparative advantages in their confrontation with states. She is also sceptical about current national and international capacities to counterbalance the main ideology of contemporary transnational terrorism—violent Islamism inspired by al-Qaeda. She stresses the quasi-religious nature of this ideology that merges radical political, social and cultural protest with the passion of belief in the possibility of a new global order.

The report argues that the mobilizing power of radical nationalism may be an alternative to transnational quasi-religious extremism at the national level. The main recommendation is that the major radical actors that combine nationalism with religious extremism be actively stimulated to further nationalize their agendas. While not a panacea, this strategy could encourage—or force—them to operate within the same frameworks as those shared by the less radical non-state actors and the states themselves.

I congratulate the author on the completion of this sharp and thought-provoking study intended for the broader public as much as for analysts and practitioners. Special thanks are also due to Dr David Cruickshank, head of the SIPRI Editorial and Publications Department, for his editing of the book, to Peter Rea for the index and to Gunnie Boman of the SIPRI Library.

Dr Bates Gill
Director, SIPRI
January 2008

Abbreviations and acronyms

CIDCM	Center for International Development and Conflict Management
ETA	Euskadi Ta Askatasuna (Basque Homeland and Freedom)
FLN	Front de libération nationale (National Liberation Front)
Hamas	Harakat al-Muqawama al-Islamiya (Islamic Resistance Movement)
IRA	Irish Republican Army
JI	Jemaah Islamiah (Islamic Group)
MIPT	Memorial Institute for the Prevention of Terrorism
PFLP	Popular Front for the Liberation of Palestine
PLO	Palestine Liberation Organization
SPIN	Segmented polycentric ideologically integrated network

1. Introduction: terrorism and asymmetry

Not all armed conflicts involve the use of terrorist means. At the same time, incidents of terrorism or even sustained terrorist campaigns can occur in the absence of open armed conflict, in an environment that would otherwise be classified as 'peacetime'. Nonetheless, in recent decades terrorism has been most commonly and systematically employed as a tactic in broader armed confrontations. However, although terrorism and armed conflict are not separate phenomena, they do not merely overlap, especially if they are carried out by the same actors.

Terrorism is integral to many contemporary conflicts and should be studied in the broader context of armed violence. The number of state-based armed conflicts gradually and significantly decreased between the early 1990s and the mid-2000s, as has the number of battle-related deaths in state-based conflicts since the 1950s.[1] However, these positive trends are counterbalanced by worrying developments and potential reversals.[2] Some of the worst trends in armed violence are related to the use of terrorism as a standard tactic in many modern armed conflicts.

First, while the numbers of state-based armed conflicts and of battle-related deaths have declined, the available data have not yet shown a comparable, major decrease in violence that is not initiated by the state—that is, in violence by non-state actors. The good news is

[1] State-based conflicts involve the state as at least one of the parties to the conflict. According to the data set of the Uppsala Conflict Data Program (UCDP) and the International Peace Research Institute, Oslo (PRIO), which covers the period since 1946, the number of armed conflicts in 2003 was 40% lower than in 1993. University of British Columbia, Human Security Centre, *Human Security Report 2005: War and Peace in the 21st Century* (Oxford University Press: New York, 2005), <http://www.humansecurityreport.info/>; and University of British Columbia, Human Security Centre, *Human Security Brief 2006* (Human Security Centre: Vancouver, 2006), <http://www.humansecuritybrief.info/>.

[2] The continuous decline in state-based conflicts since the 1990s may have stopped in the mid-2000s, as the number of such conflict remained constant at 32 for 3 years (2004–2006), following the post-cold war period low of 29 conflicts in 2003. Harbom, L. and Wallensteen, P., 'Armed conflict, 1989–2006', *Journal of Peace Research*, vol. 44, no. 5 (Sep. 2007), p. 623. Other data show that the number of states engaged in armed conflicts continues to rise and that new armed conflicts have been erupting at roughly the same pace for the past 60 years. Hewitt, J. J., Wilkenfeld, J. and Gurr, T. R., University of Maryland, Center for International Development and Conflict Management (CIDCM), *Peace and Conflict 2008* (CIDCM: College Park, Md., 2008), p. 1. Starting from the 2008 report, the CIDCM overview of trends in global conflict is also based on the UCDP–PRIO data set.

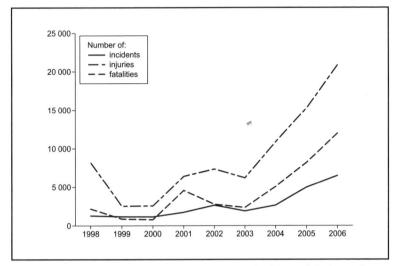

Figure 1.1. Domestic and international terrorism incidents, injuries and fatalities, 1998–2006

Source: MIPT Terrorism Knowledge Base, <http://www.tkb.org>.

that this type of violence is generally less lethal than major wars; the bad news is that it is primarily and increasingly directed against civilians.[3] Terrorism is the form of violence that most closely integrates one-sided violence against civilians with asymmetrical violent confrontation against a stronger opponent, be it a state or a group of states.

Second, in this age of information and mass communications, of critical importance is not just the scale of armed terrorist violence and its direct human and material costs, but also its destabilizing effect on national, international, human and public security and its ability to affect politics. A series of high-profile, mass-casualty terrorist attacks of the early 21st century carried out in various parts of the world demonstrate that it no longer takes hundreds of thousands of battle-related deaths to dramatically affect or destabilize international security and significantly alter the security agenda of major states and international organizations. While the number of deaths caused by the

---

[3] On patterns of violence against civilians in armed conflicts see e.g. Eck, K. and Hultman, L., 'One-sided violence against civilians in war: insights from new fatality data', *Journal of Peace Research*, vol. 44, no. 2 (Mar. 2007), pp. 233–46.

11 September 2001 terrorist attacks on the United States (almost 3000 fatalities, most of them civilians) is not comparable to the huge military and civilian death tolls of the major post-World War II wars such as those in Korea or Viet Nam, the political impact of the 2001 attacks and their repercussions for global security are comparable.

This destabilizing effect is the hallmark of terrorism and far exceeds its actual damage. It also helps explain why mere numbers do not suffice to assess the real scale, scope and political and security implications of terrorism. This characteristic makes terrorism perhaps the most asymmetrical of all forms of political violence.

Third, while many forms of armed political violence appear to be declining or stabilizing, terrorism has been clearly on the rise.[4] The year 2001 by no means marked a peak of terrorist activity over the period since 1998 (for which comprehensive data are available).[5] Since 1998, the main indicators of global terrorist activity (i.e. numbers of incidents, injuries and fatalities) have increased significantly.

The annual number of terrorist incidents—both domestic and international—rose less sharply and more steadily than the number of casualties (injuries and fatalities) over the period 1998–2006, but they still grew fivefold (rising from 1286 to 6659 attacks; see figure 1.1). Following a decline in the annual number of casualties in the late 1990s, a sharp rise caused by the high death toll of the 11 September 2001 attacks and a slight decrease in the following 18 months, casualty figures started to rise rapidly in 2003. As a result, over the period 1998–2006, the number of annual terrorism-related fatalities increased 5.6-fold (from 2172 to 12 070 fatalities), aggravated by a more than 2.6-fold increase in annual rates of terrorism-related injuries (from 8202 in 1998 to 20 991 in 2006).

---

[4] The main data set on terrorism used in this study is the MIPT Terrorism Knowledge Base, <http://www.tkb.org/>, compiled by the Memorial Institute for the Prevention of Terrorism (MIPT), Oklahoma City. It integrates data from the RAND Terrorism Chronology and the RAND–MIPT Terrorism Incident Database. Unless otherwise noted, all calculations made and graphs presented in this volume are based on the MIPT data.

[5] While the MIPT Terrorism Knowledge Base provides continuous statistical data on 'international' terrorism for the period since 1968, it only provides complete data, including statistics on 'domestic' terrorism, for the period since 1998. A first attempt to fill this gap in domestic terrorism data for the pre-1998 period is the Global Terrorism Database, which is being developed by the University of Maryland Center for International Development and Conflict Management (CIDCM) and covers both domestic and international terrorism (initially, for the period 1970–97). However, this database is likely to have a bias towards overstating the main indicators of terrorist activity as it employs too broad a definition of terrorism (that includes e.g. economically motivated acts of violence).

Not surprisingly, the most dramatic increase in terrorist activity worldwide occurred after 2001. At 6659, the number of terrorist incidents in 2006 was the largest ever recorded. This figure is a 33 per cent increase over the 4995 terrorist incidents in 2005 and a near four-fold increase since 2001 (1732 incidents). Similarly, the 2006 death toll of 12 070 showed a 47 per cent increase from the previous year and exceeded the high fatalities total for 2001 (4571 deaths) by 164 per cent.[6]

While the interim peak of terrorist activity in 2001 was primarily linked to the 11 September attacks and their immediate impact, starting from 2003, the main indicators of terrorist activity owe much of their sharp increase to the conflict in Iraq. In 2003 the 147 terrorist incidents in Iraq comprised just 8 per cent of the global total of 1899; in 2004 that share rose to 32 per cent (850 out of 2647), and in 2005 to 47 per cent (2349 out of 4995). In 2006 the conflict in Iraq accounted for a clear majority (60 per cent) of all terrorist incidents worldwide (3968 out of the global total of 6659). Similar dynamics can be traced in the growing proportion of overall terrorism-related deaths that occur in Iraq: from 23 per cent of all fatalities in 2003 (539 out of 2349) to 79 per cent in 2006 (9497 out of 12 070).[7]

As is clear from this statistical overview, one of the main stated goals of the US-led 'global war on terrorism'—to curb or diminish the terrorist threat worldwide—has largely failed. All major indicators of terrorism activity show that the overall situation has gravely deteriorated since 2001, partly as a consequence of the 'global war on terrorism' itself. A fresh look at the role of terrorism in asymmetrical conflict in needed. Before a new approach to addressing this problem can be formulated, the basic prerequisites for—and advantages of—the use of terrorism by militant non-state actors at levels from the local to the global need to be explored.

This introduction continues by proposing a new typology of terrorism, by outlining the definition of terrorism used in this report, by examining the meaning of the term 'asymmetrical conflict' and by considering the main prerequisites of terrorism in armed conflict.

---

[6] While in 2007 the numbers of terrorist attacks, fatalities and injuries decreased compared with the peak years of 2005–2006, all these indicators were still higher than the annual totals for 2001–2004. As of Jan. 2008, the data for Jan.–Nov. 2007 recorded 2747 incidents, 14 629 injuries and 6927 fatalities. MIPT Terrorism Knowledge Base (note 4).

[7] In the first 11 months of 2007 Iraq accounted for 69% of the world's terrorist incidents, 86% of fatalities and 86% of injuries. MIPT Terrorism Knowledge Base (note 4).

Chapters 2 and 3 analyse the ideological patterns of the two main forms of modern terrorism—radical nationalism and religious extremism. Chapters 4 and 5 address the organizational forms of terrorism in asymmetrical conflict at the more localized levels and the transnational level. The concluding chapter outlines strategic directions for dealing with the combination of ideologies and structures found in contemporary terrorist groups and movements.

I. Terrorism: typology and definition

Terrorism is a much debated notion. The lack of a universally recognized definition of the term is to some extent predetermined by its highly politicized, rather than purely academic, nature and origin. This allows for different interpretations depending on the purpose of the interpreter and on the political demands of the moment. However, apart from these subjective factors, there are objective reasons for the lack of agreement on a definition of terrorism—namely, the diversity and multiplicity of its forms, types and manifestations.

Traditional typologies of terrorism

This multiplicity of forms explains why the definition of terrorism cannot be separated from it typology. The two most basic, traditional and commonly used typologies of terrorism are that of domestic versus international terrorism and typology by motivation.[8] Whether these traditional classifications adequately reflect terrorism in its modern forms needs to be assessed.

A first basic distinction has traditionally been made between *domestic* and *international* terrorism. This distinction appears to have become increasingly blurred, especially if 'international terrorism' is defined as terrorist activities conducted on the territory of more than one state or involving citizens of more than one state (as victims or perpetrators). Major data sets on terrorism and the anti-terrorism legislation of many states still use this definition.[9] Few analysts and

---

[8] These traditional typologies are both widely employed for analytical and data-collection purposes. See e.g. the MIPT Terrorism Knowledge Base (note 4).

[9] E.g. according to the methodology of the RAND–MIPT Terrorism Incident Database, international terrorism is defined as 'Incidents in which terrorists go abroad to strike their targets, select domestic targets associated with a foreign state, or create an international inci-

data sets have devised more nuanced and adequate definitions of domestic terrorism.[10]

Even in the past, the distinction between international and purely domestic (home-grown or internal) terrorism was never strict and separating one from the other was not entirely accurate, because the two have always been intimately interconnected. Terrorist activity, especially when perpetrated on a regular, systematic basis, was rarely fully self-sufficient and contained within the borders of one state. The internationalist ideology of a terrorist group often required it to extend its actions beyond a national context (as exemplified by the assassinations of leaders of several European states by Italian anarchists in the late 19th century). In addition, terrorists have often had to internationalize financial, technical, propaganda and other aspects of their activity. For instance, in the early 1900s Russian Socialist Revolutionaries (the SRs or Esers) found refuge, planned terrorist attacks and produced explosives in France and Switzerland. Terrorism employed by anti-colonial and other national liberation movements in the late 19th and 20th centuries (e.g. against British rule in India) was internationalized *de facto*, if not *de jure*. The high degree of internationalization was also one of the main characteristics of leftist terrorism in Western Europe and elsewhere in the 1970s and 1980s, when terrorists from several European states mounted joint operations or trained together, for example in Palestinian training camps in the Middle East. At this time Japanese Red Army members were frequently relocating from one country to another.

By the end of the 20th century, the distinction between domestic and international terrorism had become more blurred than ever.[11]

dent by attacking airline passengers, personnel or equipment'. Domestic terrorism is defined as 'Incidents perpetrated by local nationals against a purely domestic target'. MIPT Terrorism Knowledge Base, 'TBK: data methodologies: RAND Terrorism Chronology 1968–1997 and RAND–MIPT Terrorism Incident database (1998–present)', <http://www.tkb.org/Rand Summary.jsp?page=method>. The US legislation defines international terrorism as 'terrorism involving citizens or the territory of more than one country'. United States Code, Title 22, Section 2656f(d).

[10] E.g. according to the Terrorism in Western Europe: Event Data (TWEED) data set methodology, terrorism is internal when terrorists originate and act within their own political systems. See Engene, J. O., 'Five decades of terrorism in Europe: the TWEED dataset', *Journal of Peace Research*, vol. 44, no. 1 (Jan. 2007), pp. 109–10.

[11] Against this background, it is not surprising that Europol, the European Police Office, has decided to no longer use the distinction between domestic and international terrorism in its analytical assessments of the terrorist threat. Europol, *EU Terrorism Situation and Trend Report 2007* (Europol: The Hague, 2007), <http://www.europol.europa.eu/index.asp?page= publications>, p. 10.

Those terrorist groups whose political agenda remained localized to a certain political or national context tended to increasingly internationalize some or most of their logistics, fund-raising, propaganda and even planning activities, sometimes extending them to regions far from their main areas of operation. Even terrorist groups with localized goals are now likely to be partly based and operate from abroad.[12] In fact, in the modern world there are few groups that have employed terrorist tactics that rely on domestic resources and means alone. Groups engaged in armed conflicts in very remote locations (e.g. the Maoists in Nepal) who relied primarily on internal resources still build ideological links with like-minded movements (in the case of the Nepalese Maoists, with the Naxalite movement in India, among others) and obtained some financial or logistical support from abroad. In a peacetime environment, acts of purely domestic terrorism are usually limited to isolated terrorist attacks by left- or right-wing extremists (e.g. the April 1995 Oklahoma City bombing in the USA).

It should be stressed that the high degree of internationalization of terrorist activities by both communist and other leftist groups and the more recent violent Islamist networks has been rarely driven by pragmatic logistical needs alone. It is also a natural progression of their internationalist (transnationalist, supranational) ideologies and world views. Thus, for instance, some of the semi- or fully autonomous Islamist terrorist cells in Europe, comprising radical Muslims who may be citizens of European states, may have limited—or no—direct operational guidance, financial support or other logistical links with the rest of the transnational violent Islamist movement. However, these cells' terrorist activities should still be viewed as manifestations of transnational terrorism as long as they are guided by a universalist, quasi-religious ideology and are carried out in the name of the entire *umma* and in reaction to Western interventions in Afghanistan, Iraq or elsewhere.[13] This is transnational terrorism, even if it results in civilian casualties primarily among the perpetrators' fellow citizens.

It is important to distinguish between the different forms, levels and stages of the gradual erosion of a strict divide between international and domestic terrorism. The erosion may, for instance, be limited to a

[12] E.g. the Liberation Tigers of Tamil Eelam (LTTE), whose political goals do not go beyond intra-state, ethno-political conflict in Sri Lanka, have one of the most widespread logistics and support networks in the world.

[13] The term *umma* is mainly used here to mean the entire Muslim world or community. On the meaning of the term 'quasi-religious' see chapter 3 in this volume.

simple internationalization of a terrorist group's activities: conducting terrorist acts abroad or extending logistics and fund-raising activities to foreign countries. It may also take a more advanced form of transnationalization: ranging from more active interaction between independent groups in different countries to the formation of fully fledged inter-organizational networks or even, ultimately, to the emergence of transnational terrorist networks. In sum, of primary importance today is not the mechanical distinction between domestic and international terrorism, but whether a group's overall goals and agenda are confined to the local and national levels or are truly transnational or even global. In this Research Report, the term 'internationalized' is applied to terrorism and groups engaged in terrorist activity at levels from the local to the regional that prioritize goals within a national context. The term 'transnational' is reserved for terrorist networks operating and advancing an agenda at an inter-regional or even global level.

The second traditional typology of terrorism addressed here is based on a group's *dominant motivation*. According to this criterion, terrorist groups are normally allocated to one of three broad categories: (*a*) socio-political (or secular ideological) terrorism of a revolutionary leftist, anarchist, right-wing or other bent; (*b*) nationalist terrorism, ranging from that practised by national liberation movements fighting colonial or foreign occupation to that employed by ethno-separatist organizations against central governments; and (*c*) religious terrorism, practised by groups ranging from totalitarian sects and cults to broader movements whose ideology is dominated by religious imperatives.

Since the early 1990s, following the end of the cold war, international terrorist activity by socio-political, particularly communist or leftist groups, has understandably declined, in terms of both incidents and casualties (see figures 2.1 and 2.2 in chapter 2). While the combined dynamics of international and domestic terrorism of this type have shown some increase in absolute numbers since 1998 (see figure 2.3 below), in relative terms, terrorist activities by communist or leftist groups have been conducted at a lower level than those by nationalist and religious groups, especially in terms of casualties. The annual global totals of injuries and fatalities due to communist or leftist groups number hundreds, as compared to thousands for the other two motivational types. In contrast, the overall dynamics of nationalist and, especially, religious (mostly Islamist) terrorism, while highly uneven, indicate that terrorism of both types has been sharply rising in both absolute and relative terms, particularly since the late 1990s.

The main problem with the motivational typology is that in practice few groups have a 'pure' motivation formulated in accordance with its ideology. Many militant–terrorist groups are driven by more than one motivation (and more than one ideology).[14] It may not always be clear which motivation is dominant; one motivation may replace another with time or they can gradually merge. Some of the most common combinations have included: (*a*) a synthesis of right-wing extremism and religious fundamentalism; (*b*) a mix of nationalism and left-wing radicalism; and (*c*) religious extremism merged with radical national-ism (e.g. the Palestinian groups Hamas and Islamic Jihad and the Islamicized nationalist groups of the Iraqi resistance movement) or with ethno-separatism (e.g. the Sikh, Kashmiri and Chechen separat-ists). Thus, while motivational typology remains important, it does not always adequately and accurately reflect the complex, dialectic nature of terrorist groups' motivations and ideologies.

The functional typology of terrorism

The need to revise and supplement traditional typologies of terrorism has led the present author to suggest what may be called the 'func-tional' typology of terrorism. It is centred on the function that terrorist tactics play for a non-state actor depending on its level of activity and in relation to an armed conflict. Consequently, this typology is based on two criteria: (*a*) the level and scale of a group's ultimate goals and agenda (i.e. whether global or more localized); and (*b*) the extent to which terrorist activities are related to or are part of a broader armed confrontation and are combined with other forms of armed violence.

On the basis of these two criteria, three functional types of modern terrorism can be distinguished.

1. *The 'classic' terrorism of peacetime.* Examples of this include communist and other leftist terrorism in Western Europe in the 1970s and the 1980s; right-wing terrorism when it is not a tactic used by loyalist and other anti-insurgency groups in armed conflict; and eco-

---

[14] The term 'militant–terrorist' is used in this study to refer to militant groups that employ terrorist means alongside other violent tactics. In most conflict settings, it is a more accurate term than either 'militant' or 'terrorist'. On groups using more that one violent tactic see below.

logical or other special interest terrorism. Regardless of its motivation, terrorism of this type is independent of any broader armed conflict and, as such, is not a subject of this Research Report.[15]

2. *Conflict-related terrorism.* Such terrorism is systematically employed as a tactic in asymmetrical local or regional armed conflicts (e.g. by Chechen, Kashmiri, Palestinian, Tamil and other militants). Conflict-related terrorism is tied to the concrete agenda of a particular armed conflict and terrorists identify themselves with a particular political cause (or causes)—the incompatibility over which the conflict is fought. This cause may be quite ambitious (e.g. to seize power in a state, to create a new state or to fight against foreign occupation), but it normally does not extend beyond a local or regional context. In this sense, the terrorists' goals are limited, as are the technical means they normally use. Conflict-related terrorism is practised by groups that enjoy at least some local popular support and tend to use more than one form of violence. For example, they frequently combine terrorist means with guerrilla attacks against regular army and other security targets or with symmetrical inter-communal, sectarian and other violence against other non-state actors.

3. *Superterrorism.* While the other two types of terrorism are more traditional, superterrorism is a relatively new phenomenon (also known as mega-terrorism, macro-terrorism or global terrorism).[16] Superterrorism is by definition global or at least seeks global outreach and, as such, does not have to be tied to any particular local or national context or armed conflict. Superterrorism ultimately pursues existential, non-negotiable, global and in this sense unlimited goals— such as that of challenging and changing the entire world order, as in the case of al-Qaeda and the broader, post-al-Qaeda transnational violent Islamist movement.[17]

[15] See also note 51.
[16] See e.g. Freedman, L. (ed.), *Superterrorism: Policy Responses* (Blackwell: Oxford, 2002); and Fedorov, A. V. (ed.), *Superterrorizm: novyi vyzov novogo veka* [Superterrorism: a new challenge of the new century] (Prava Cheloveka: Moscow, 2002). Prior to 11 Sep. 2001, the term superterrorism was primarily used as a synonym for terrorism employing unconventional (chemical, biological, radiological and nuclear) means. In contrast, in this Research Report, the ultimate level of the goals, rather than the nature of the technical means employed, serves as the main criteria for defining this type of terrorism.
[17] In this study, the term 'post-al-Qaeda' refers to the broader transnational violent Islamist movement that evolved after the 11 Sep. 2001 attacks on the USA, was inspired and instigated by the original al-Qaeda but represents a different—and dynamic—type of organization. While the term 'post-al-Qaeda' points towards the original al-Qaeda as the main inspirer

While these three types of terrorism are functionally different and retain specific features of their own, they share some characteristics, may be interconnected, interact and, in some cases, even merge. For example, home-grown conflict-related terrorist activity in the armed conflicts in Afghanistan or Iraq can be inspired by the actions of cells of transnational superterrorist networks and can adopt or imitate their tactics, and vice versa.

Despite the emergence and rise of superterrorism and the fact that it dominates anti-terrorism agendas in the West, especially since the unprecedented superterrorist attacks of 11 September 2001, terrorism systematically employed as a tactic in local or regional asymmetrical armed conflicts remains the most widespread form of terrorism. This is the most basic and common form of modern terrorism and it continues to result in the largest total numbers of terrorist incidents and terrorism-related injuries and fatalities. At the end of the 19th century, the same role was played by left-wing, revolutionary terrorism, which was mostly carried out in otherwise peacetime settings.[18] It remains to be seen if superterrorism, with its transnational, global outreach and agenda, will assume that role in the not too distant future.

The main definitional criteria of terrorism

The definition of terrorism used in this Research Report is the intentional use or threat to use violence against civilians and non-combatants by a non-state (trans- or sub-national) actor in an asymmetrical confrontation, in order to achieve political goals.[19]

This definition narrows the scope of activities in the category of 'terrorism' to the maximum possible extent. At least three main criteria may be used to distinguish terrorism from the other forms of violence with which it is often confused, especially in the context of a

and the ideological and organizational origin of this much broader movement, it more accurately reflects the fact that the movement is no longer confined to the jihadi veteran networks that emerged in the course of the anti-Soviet jihad in Afghanistan and formed the core of al-Qaeda. Structurally, this broader movement represents a new type of organization; see chapter 5 in this volume. On the ideology of the movement see chapter 3 in this volume. This movement is also frequently referred to, particularly in Western literature, as 'global jihad', 'global Salafi jihad' or 'the jihadi–Salafi current of global jihad'.

[18] Exceptions are the few cases where it was employed as one of several violent tactics in revolutions or revolts (such as the first Russian revolution of 1905–1907).

[19] In terrorist incidents, civilians may be specifically targeted or they may be the inevitable victims of indiscriminate violence.

broader armed confrontation. If a certain act or threat of violence fits all three criteria, it can be characterized as a terrorist act.[20]

The first criterion—*a political goal*—distinguishes terrorism from crime that is motivated by economic gain, including organized crime.[21] The political goal can range from the concrete to the abstract. While such a goal may include ideological or religious motivations or be formulated in ideological or religious terms, it always has a political dimension. For groups engaged in terrorism, a political goal is an end in itself, not a secondary instrument or a cover for advancement of other interests, such as illegal accumulation of wealth. Terrorists may imitate or employ criminal means of generating money for self-financing and may interact with organized crime for the same aim. However, whereas for criminals gaining the greatest material profit is the ultimate goal, for terrorists it is primarily the means to advance their main political, religious or ideological goals. In some cases terrorist attacks may be partly motivated by economic gain, but this is not these groups' sole or dominant *raison d'être*.

It should also be stressed that terrorism is not the political goal itself, but a specific tactic to achieve that goal (thus, it makes sense to refer to 'terrorist means', rather than 'terrorist goals'). Different groups may have the same political goal but may use different forms of violence, combine different tactics and even use non-violent means to achieve that goal. The important implication is that if a group chooses terrorism as a means to achieve a political goal, the aim of its struggle, however benign, cannot be used to justify its actions. However, the fact that a group uses terrorist means in the name of a political goal does not necessarily delegitimize the goal itself.

The second criterion—*civilians as the direct target of violence*—helps distinguish terrorism from some other forms of politically motivated violence, particularly those used in the course of armed conflicts. The most notable of these is guerrilla warfare, which implies the use of force against governmental military and security forces by

---

[20] On definitional issues see Stepanova, E., *Anti-terrorism and Peace-building During and After Conflict*, SIPRI Policy Paper no. 2 (SIPRI: Stockholm, June 2003), <http://books.sipri.org/>, pp. 3–8; and Stepanova, E., 'Terrorism as a tactic of spoilers in peace processes', eds E. Newmann and O. Richards, *Challenges to Peacebuilding: Managing Spoilers during Conflict Resolution* (United Nations University Press: Tokyo, 2006), pp. 83–89.

[21] This has been noted as a defining characteristic of terrorism by many, if not most, scholars who had specialized in terrorism studies before and after 11 Sep. 2001. E.g. most notably in Hoffman, B., *Inside Terrorism*, revised edn (Columbia University Press: New York, 2006), pp. 2, 40.

the rebels who presumably enjoy the support of at least part of the local population in whose name they claim to fight. In contrast, terrorism is specifically directed against the civilian population and civilian objects or is intentionally indiscriminate. This does not mean that a certain armed movement cannot simultaneously use different modes of operation, including both guerrilla and terrorist tactics, or switch between these tactics. Accordingly, this Research Report uses such terms as 'militant–terrorist groups', 'organizations involved in terrorist activities' or 'groups using terrorist means', rather than 'terrorist organizations', for groups that use more than one violent tactic.

This criterion is not absolute, as in some cases it might be difficult to identify a target as civilian, to prove that civilians were intentionally targeted or to distinguish between combatants and non-combatants in a conflict area. However, it is still useful. The target of violence also has serious implications in international humanitarian law. Guerrilla attacks against government military and security targets are not internationally criminalized (although domestically they usually are). However, deliberate attacks against civilians committed in the context of either inter- or intra-state armed conflict, including terrorist attacks, are direct violations of international humanitarian law.[22]

While terrorism is a specific tactic that necessitates victims, and while civilians remain the most immediate targets of terrorism, those victims are not the intended end recipients of the terrorists' message. Terrorism is a performance that involves the use or threat to use violence against civilians, but which is staged specifically for someone else to watch. Most commonly, the intended audience is a state (or a group or community of states) and the terrorist act is meant to blackmail the state into doing or abstaining from doing something. The state as the ultimate recipient of the terrorists' message leads to the third defining criterion—*the asymmetrical nature of terrorism*.

There are several forms of politically motivated violence against civilians, particularly in the context of an ongoing armed conflict.

---

[22] '[T]he Parties to the conflict shall at all times distinguish between the civilian population and combatants and between civilian objects and military objectives and accordingly shall direct their operations only against military objectives'. Article 48 of the Protocol Additional to the 1949 Geneva Conventions, and relating to the Protection of Victims of International Armed Conflicts (Protocol I), opened for signature on 12 Dec. 1977 and entered into force on 7 Dec. 1978. The international law regulating non-international armed conflict (Protocol II) does not prohibit members of rebel forces from using force against government soldiers or property provided that the basic tenets governing such use of force are respected. The texts of the 2 protocols are available at <http://www.icrc.org/ihl.nsf/CONVPRES>.

Repressive actions by the state against its own or foreign civilians or symmetrical inter-communal violence on an ethnic, sectarian or other basis may also meet the first two criteria mentioned above. What distinguishes terrorist activity from these and some other forms of politically motivated violence against civilians and non-combatants is the asymmetrical aspect of terrorism. It is used as a weapon of 'the weak' against 'the strong'. Furthermore, it is a tactic of the side that is not only physically and technically weaker but also has a lower formal status in an asymmetrical confrontation ('status asymmetry').[23]

It is the asymmetrical nature of terrorism that explains the terrorists' perceived need to attack civilians or non-combatants. They perceive it as serving as a force multiplier that compensates for conventional military weakness and as a public relations tool to exert pressure on the state and society at large. A terrorist group tries to strike at the strong where it hurts most, by mounting or threatening attacks against civilians and civil infrastructure. Terrorism is a weapon of the weak (non-state actors) to be employed against the strong (states and groups of states). It is neither a weapon of the weak to be symmetrically employed against the weak, nor a weapon of the strong.[24]

II. Asymmetry and asymmetrical conflict

One of the implications of the asymmetrical nature of terrorism is that it cannot be employed as a mode of operation in all armed conflicts. It is used only in those conflicts that have some asymmetrical aspect.

Asymmetry in armed conflict has been most often interpreted as a wide disparity between the parties, primarily in military and economic

---

[23] On status asymmetry see below.

[24] Repressive actions and deliberate use of force by the state against its own or foreign civilians and non-combatants are not included in the definition of terrorism used in this study because they are not applied by a weaker actor of a lower status in an asymmetrical armed confrontation. This definition does not prevent the use of the term 'terror' (instead of terrorism) to describe state repression. Nor does it exclude state support to non-state (trans- or subnational) groups engaged in terrorist activity. However, in cases where this support amounts to or transforms into full and direct control and strategic guidance over a clandestine group, it makes sense to refer to this group's activities as being 'covert', 'secret', 'sabotage' or other state-directed operations in the classic sense rather than terrorism as such. The need to internationally criminalize those repressive actions against civilians that are committed by states on a massive scale in a situation short of armed conflict of either international or non-international nature (and are thus not covered by the international humanitarian law, protocols I and II (note 22)) is still pressing. However, this is not a sufficient reason to extend the notion of terrorism to cover these actions.

power, potential and resources. As well as being overly militarized, this approach is both too broad and too narrow to adequately describe the nature of terrorism in asymmetrical conflicts.

Demilitarizing asymmetry

The standard and in many ways outdated definition of asymmetry in armed conflict is narrowed by its excessively militarized nature. However, it is still broad enough to suggest that most armed conflicts worldwide are fully or partly asymmetrical, with the exception of the few symmetrical interstate confrontations (i.e. conflicts between regional powers with relatively similar military and economic potential, such as the 1980–88 Iran–Iraq War) or conflicts between non-state actors. Such a broad definition encompasses a wide spectrum of armed confrontations. At one end of this spectrum are internal conflicts between a state and a sub- or non-state opponent at home or abroad. At the other are conflicts between states with radically different levels of military and economic potential, most of which take the form of military interventions of the incomparably 'stronger' side against the 'weaker' one. According to this approach, the absolute military–technological superiority of the USA over any other actual or potential opponent means that nearly every armed conflict in which the USA may be engaged is by definition asymmetrical. At the interstate level, recent examples of asymmetric conflict include the US-led military interventions in Iraq in 1991 and 2003. It is not surprising that, within this militarized framework, the term 'asymmetrical warfare' is preferred to 'asymmetrical conflict'. It is used to denote a military tactic (or mode of operation) that exploits the opponent's weaknesses and vulnerabilities and emphasizes differences in forces, technologies, weapons and rules of engagement.[25]

This view is one-sided in its military focus and strikingly straightforward in its vagueness. However, this does not mean that Western military or politico-military thought has not generated anything more nuanced and better tailored to the main type of contemporary armed conflict—intra-state conflicts that may be internationalized to a varying extent—and the threats that it poses. It suffices to mention that

---

[25] US Department of the Army, Headquarters, *Operational Terms and Symbols*, Field Manual no. 1-02/Marine Corps Reference Publication no. 5-2A (Department of the Army: Washington, DC, 2002), p. 21.

even before the end of the cold war, the USA was the only state that had at its disposal a doctrine for participation in sub-conventional, or 'low intensity', conflicts. That doctrine emerged in the wake of the USA's military failure in Viet Nam (1965–73) and reflected the type of conflict in which the USA found itself increasingly involved during the last decade of the cold war.[26] These conflicts appeared to be quite different from conventional interstate wars of medium intensity and were far short of a high-intensity global confrontation involving the use of nuclear arms. The strategy for fighting low-intensity conflicts was both well developed in doctrinal terms and applied by the USA in practice (e.g. in El Salvador).

For this Research Report, of special interest is not so much the intensity aspect of this theory as the growing attention it paid to the asymmetrical character of the forms of violence most typical for these conflicts (i.e. insurgency, terrorism etc.). Of particular importance is the limited but remarkable extent to which the USA's low-intensity conflict doctrine went beyond a purely military outlook in interpreting the nature of asymmetry in conflict. Among other things, this theory was the first of its kind in the post-World War II period to focus on the protagonists' different political and psychological capacity to accept human losses. It also noted the moral superiority of an 'enemy' which is otherwise incomparably weaker in the conventional (military, technological and economic) sense. The doctrine was the first attempt to combine the political, economic, information and military tools required for an asymmetrical low-intensity confrontation of this type.

In the following decades of the late 20th and early 21st centuries some US military analysts effectively developed and revised this tradition within various conceptual frameworks. They insisted on the need to extend the notion of asymmetry from just acting differently to 'organizing, and thinking differently than opponents' and for the term to imply not just standard differences in methods and technologies,

---

[26] For the doctrinal principles and specifics of US participation in asymmetrical, 'low-intensity' conflicts see e.g. US Department of the Army, Headquarters, *Low-Intensity Conflict*, Field Manual no. 100-20 (Government Printing Office: Washington, DC, 1981). For an updated version see US Department of the Army, Headquarters, *Operations in a Low-Intensity Conflict*, Field Manual no. 7-98 (Government Printing Office: Washington, DC, 1992).

but also disparities in 'values, organizations, time perspectives'.[27] Some of the most up-to-date and advanced counter-insurgency military doctrines—strategic thinking that is by default required to prioritize threats from opponents taking asymmetrical approaches—describe terrorist and guerrilla attacks employed by insurgents as asymmetrical threats 'by nature', 'planned to achieve the greatest political and informational impact' and requiring commanders to understand how a non-state opponent 'uses violence to achieve its goals and how violent actions are linked to political and informational operations'.[28]

With the wide and quick proliferation of asymmetrical threats, the need to further demilitarize the definition and understanding of asymmetry in conflict has become more urgent than ever. This Research Report uses the terms 'asymmetrical confrontation' and 'asymmetrical conflict', rather than the term 'asymmetrical warfare'. The latter term is a narrow one because it is still mainly defined by military power criteria. It is also an excessively broad one to the extent that it applies to conflicts between states, conflicts within states and conflicts that go beyond state borders but involve actors of different statuses. Indeed, the notion of 'asymmetrical confrontation' should be further extended to go beyond the gaps in military potential or military power. Counter-intuitively, this is exactly what permits the limiting and narrowing down of the range of conflicts that this term may be applied to, primarily due to the different 'status' characteristics of the main protagonists.

Power asymmetry

So-called power asymmetry is the core component of most traditional—and excessively militarized—definitions of asymmetry in conflict. It remains an important component of the definition of asymmetrical conflict used here. It is particularly relevant in view of

[27] Metz, S. and Johnson, D. V., *Asymmetry and U.S. Military Strategy: Definition, Background, and Strategic Concepts* (US Army War College, Strategic Studies Institute: Carlisle, Pa., Jan. 2001), pp. 5–6. For a discussion of this broader version of asymmetry see also Reynolds, J. W., *Deterring and Responding to Asymmetrical Threats* (US Army Command and General Staff College, School of Advanced Military Studies: Fort Leavenworth, Kans., 2003).

[28] US Department of the Army, Headquarters, *Counterinsurgency*, Field Manual no. 3-24/ Marine Corps Warfighting Publication no. 3-33.5 (Department of the Army: Washington, DC, Dec. 2006), p. 3-18.

the terrorists' need for a form of violence that serves as a force multiplier in confrontation with an incomparably stronger opponent that they cannot effectively challenge by conventional means. This need conditions the terrorist mode of operation that attacks the enemy's weakest points: its civilians and non-combatants. However, the power gap should be viewed as only one of the two essential characteristics that favour the conventionally stronger side and, overall, just one of four key characteristics of a two-way asymmetry (discussed in the following sections).

Three additional points in relation to power asymmetry between the parties are often overlooked.

First, the power disparities discussed here are not marginal or relative, but extreme. This is the case even if the interpretation of the notion of 'power' is not extended indefinitely to embrace all spheres of life and is sufficiently well covered by focusing on conventional (i.e. economic, military and technological) aspects.

Second, the extreme imbalance in resources available to parties to an asymmetrical confrontation is partly, although not decisively, compensated for by the reverse imbalance in resources that each side needs in order to effectively confront the opponent. In other words, terrorism always requires far fewer financial, technical and other conventional resources than counterterrorism.

Third, the enormously higher power resources of the stronger side in an asymmetrical conflict by definition lead to asymmetrically high conventional damage and high numbers of victims for its opponents. In other words, the weaker side always suffers incomparably higher total conventional losses in an armed conflict (both battle-related and civilian). Of all asymmetrical ways to strike back that are available to a weaker party, terrorism is perhaps the most effective way to balance this asymmetry by making enemy civilians suffer as much as those in whose name the terrorist claim to act.

Status asymmetry

As noted above, most definitions of asymmetrical conflict prioritize 'power' disparities based on quantifiable parameters (military budgets, weapons arsenals, technological superiority etc.). To these some may add other, mainly politico-military, dimensions of power, such as asymmetry of purpose or a sharp contrast between the two sides in their overall understanding and interpretation of security.

The first step needed to go beyond the 'power' factor is to recognize that asymmetry has a qualitative, as well as a quantitative, dimension. The best way to embrace most of the non-quantifiable aspects of power is to introduce an additional qualitative criterion—the party's formal status in the existing system, at both the national and the international levels. In other words, the conflict is fully asymmetrical when the notion of power is extended to include a status imbalance, that is, when the conflict is between actors of different status. The most basic form of such conflict is a confrontation between a non-state actor and a state, or states.[29]

This double asymmetry (power plus status) has the additional advantage of limiting the range of actual armed conflicts studied to those where terrorism can be employed as a tactic of non-state actors. Adding the status dimension to the notion of asymmetrical conflict does not mean that such a conflict has to be confined within the borders of one state. Nor does it mean that a non-state actor is necessarily a sub-state one. In this context, a non-state actor may well be a transnational non-state network with a global outreach. However, its confrontation with a group or community of states would still qualify as asymmetrical in terms of the gap in the protagonists' formal status within the international system as well as in terms of the traditional interpretation of power as primarily military power.

Conventional power and formal status remain the key asymmetrical assets of the state, even though both these assets may be slowly eroding—for some states more than for others—in the modern world. In this Research Report, an asymmetrical conflict is treated as conflict in which extreme imbalance of military, economic and technological power is supplemented and aggravated by status inequality; specifically, the inequality between a non- or sub-state actor and a state.

---

[29] One of several reasons why the status dimension has not been emphasized or has been ignored in much of the military and security thinking on asymmetrical threats such as terrorism (especially 'ideological', or 'socio-political' terrorism) was that for a long time, especially during the last decades of the cold war, this threat was often viewed primarily as a state-sponsored activity and was not fully recognized as a non-state phenomenon. In contrast, most contemporary definitions view terrorism as an activity that may get some state support but is not initiated by a state and is essentially a tactic employed by increasingly autonomous non-state actors.

Two-way asymmetry

Asymmetry in conflict is not just, and not even mainly, about the stronger side making use of its advantages. The asymmetry does not work in just one direction. If that were the case, then the stronger side could easily use its superior military force, technology and economic potential to decisively crush its weaker opponent.

However, alongside its multiple superiorities, a conventionally stronger side has its own inherent, organic, generic vulnerabilities that are often inevitable by-products of its main strengths and are not minor, temporary flaws that can be quickly fixed. It is these objective weaknesses that allow a conventionally weaker opponent that enjoys a lower formal status to turn a direct, top-down one-way asymmetry into a two-way one which includes a reverse, bottom-up asymmetry.

In this kind of asymmetry, the protagonists differ in their strengths and weaknesses. A common way to address the two-way nature of asymmetry has been to make a distinction between positive asymmetry (the use of superior resources by the conventionally stronger side) and negative asymmetry (the resources that a weaker opponent can use to exploit the protagonist's vulnerabilities). In this context, both power and status criteria are positive or, on a vertical scale, top-down advantages of the state. What then are the weaker side's reverse, bottom-up advantages that could qualify as negative asymmetry?

Unable to effectively fight on the enemy's own ground and to challenge a stronger opponent on equal terms, the weaker, lower status side has to find some other ground and to rely on other resources to establish a two-way asymmetry. It is important to stress that the specific strengths of the weaker party cannot be simplified, as is often done with the militarized interpretation of asymmetry, to a mere reaction and conscious, opportunistic exploitation of the opponent's vulnerabilities. This approach fails to recognize that the conventionally weaker non-state actor may also have genuine advantages and strengths that, even if they are not as easily quantifiable, are not just reactive in nature and cannot be reduced to a distorted mirror image of the stronger party.

Ideological disparity

The first advantage that anti-state armed actors, especially those that systematically employ terrorist means, have at their disposal is the very high power of mobilization and indoctrination that their radical, extremist ideologies have in certain segments of society. These ideologies, and the specific goals and agendas formulated in line with them, will have their greatest power in parts of the ethnic or religious community, social group or class in whose name the militant–terrorist actors claim to speak and whose interests they claim to defend. In other words, if there is one area where a reverse asymmetry strongly favours the weak, it is the ideological front. As summed up by Carlos Marighella, a Brazilian theorist and practitioner of 'urban guerrilla' warfare, the conventionally weaker side's 'arms are inferior to the enemy's', but 'from a moral point of view' the former enjoys 'an undeniable superiority'.[30]

That does not imply that the radical ideologies of non-state actors ready to take arms or to employ terrorist means are superior or more powerful than the mainstream ideologies of nation states or those of other, less radical non-state actors. On the contrary, the more radical an ideology is, the more utopian and unrealistic is its vision of the present and especially of the future world. However, precisely because of its radical nature, an anti-system ideology has a massive comparative advantage over any moderate one as a mobilization and indoctrination force in specific circumstances and in a specific framework (i.e. in the framework of asymmetrical confrontation at the localized or trans-national levels). The forces and actors ready to take up arms to oppose the dominant system (the political, social, national or international order) are by definition far more ideologically zealous, more strongly motivated and display a much higher level of resolve and commitment to their ideological goals than their mainstream opponents.

As argued in this Research Report, bottom-up, reverse ideological asymmetry is a key characteristic of the systematic use of terrorist means in an asymmetrical confrontation.[31] It is just as important an

---

[30] Marighella, C., *Minimanual of the Urban Guerrilla* (Paladin Press: Boulder, Colo., 1975), p. 5. The text is also available at <http://www.marxists.org/archive/marighella-carlos/1969/06/minimanual-urban-guerrilla/>.

[31] This is not to be confused with the *ideological asymmetry hypothesis* as a specific feature of social dominance theory. This latter hypothesis asserts that the relationship between attitudes towards hierarchy-maintaining social practices and anti-egalitarian social values is

element of this asymmetry as the top-down power and status advantages of the conventionally stronger side. Sharp ideological disparity is the main condition for turning what may seem a one-way asymmetry into a two-way one. It is also the basis for a host of other qualitative imbalances and dissimilarities, such as the disparities in purpose and in understanding and interpretation of 'security', 'victory', 'defeat' and so on.

Structural disparity

While such a radical ideological disparity is a *sine qua non* for a two-way asymmetrical confrontation, structural disparity—sharp dissimilarities in organizational forms and patterns employed by protagonists—although significant, is not essential as long as the three other criteria (of power, status and ideology) are met. The structural, or organizational, patterns of militant non-state actors challenging the status quo and the extent to which they may or may not imitate the organizational forms typical of their main opponent vary significantly. These patterns range from the strict hierarchies of apocalyptic religious cults to extremely loose networks of semi- or fully autonomous cells directed by general ideological and strategic guidelines from several leaders.

Against this background, two things must be stressed. First, special attention should be paid to the extent to which the radical ideology of an armed non-state actor dictates and shapes its organizational forms. Second, while the organizational patterns of militant–terrorist groups may vary, the basic assumption is that the more different these structures are from those most typical of their main protagonist (the state), the harder it is to counter the respective non-state actors in an asymmetrical confrontation.

Overall, demilitarizing the notion of asymmetry both allows the broadening of its interpretation to include disparities in formal political status, ideologies and possibly organizational patterns, and suggests a more focused definition of asymmetry in armed conflict. This definition implies a two-way asymmetry where the state has superior power and enjoys a higher formal status while a non-state actor possesses certain ideological advantages that may also be reinforced by

more positive among members of high-status groups than among members of low-status groups.

structural disparities. This definition appears to be comprehensive and better tailored for the specific purposes of this Research Report because it takes into account asymmetry in all respects and from all sides.

Finally, it should be kept in mind that not all asymmetrical threats are related to or generated by armed conflicts. Threats posed by organized crime groups, especially transnational organized crime, are also commonly characterized as 'asymmetrical'. Nor is terrorism the only method employed by weaker opponents in asymmetrical confrontation: insurgency or guerrilla warfare remains the most common of the other asymmetrical tactics. The main difference between terrorism and insurgency, in terms of their asymmetrical nature, is that terrorism is an even more unconventional and asymmetrical form of violence that produces an effective and deadly combination: one-sided violence against unarmed civilians employed against a conventionally much stronger opponent that also enjoys a higher formal status.

III. Ideological and structural prerequisites for terrorism

Of the three types of terrorism identified in line with the functional typology proposed in this Research Report, the one most directly connected to violent conflict is conflict-related terrorism. Any search for the fundamental, political, socio-economic and other drivers—the root causes[32]—of this type of terrorism inevitably boils down to analysis of the basic causes of violent conflict as such. In this context, conflict-related terrorism is just one specific tactic of violence, secondary to the broader phenomenon of armed conflict itself. It is not surprising then that the underlying, 'structural' causes of terrorism as a mode of operation in a violent conflict are generally the same as the underlying causes of the armed conflict as a whole.[33] It would be easy to conclude from this, in a rather simplistic way, that in order to effectively counter terrorism generated by, related to and used as a tactic in an asymmetrical conflict, it is not only necessary, but also sufficient to

[32] For a comprehensive and critical discussion of the notion of 'root causes' as applied to terrorism see e.g. Bjørgo, T., 'Introduction', ed. T. Bjørgo, *Root Causes of Terrorism: Myths, Reality and Ways Forward* (Routledge: Abingdon, 2005), pp. 1–6.

[33] This does not apply to the other types of terrorism identified by the functional typology—terrorism of peacetime and transnational superterrorism.

address the fundamental causes of the conflict itself and to solve or reconcile the basic incompatibilities between the parties.

The structural causes (such as incomplete, particularly uneven and 'traumatic' modernization[34]) and their more concrete manifestations (i.e. the main incompatibilities between the parties to an armed conflict) may well help to explain why the conflict has become violent. However, they do not suffice to explain why in a particular conflict-related context the violence takes the form of terrorism. The mere fact of an asymmetrical armed confrontation between a non-state actor and a state or states does not automatically imply the use of terrorist means, as bluntly demonstrated by the 2006 conflict between Israel and Lebanon-based Hezbollah. Even when terrorism is employed as a tactic in an asymmetrical confrontation, not all armed non-state groups operating in the same conflict necessarily resort to terrorist means.

In addition, when applied to the link between armed conflict and terrorism, the 'root causes' approach alone may be too static to grasp the dynamic nature of conflict itself and of terrorism used as a tactic in that conflict. Looked at from a more actor-centred perspective, over time terrorist means may start to be used by violent non-state actors for purposes other than initially planned. Their use may extend beyond the main incompatibility with the state, or they may even develop a momentum of their own and cease to remain just a function of the armed conflict. A group may also feel a growing need to resort to increasingly asymmetrical forms of violence towards the end of a conflict, as the range of other options for resistance becomes limited due to harsh suppression by a state opponent or to a peace process gaining momentum.

In sum, in addition to the fundamental root causes of violent conflict, there must be some more specific prerequisites for a non-state actor to resort to terrorism. While not necessarily as broad as the root causes of the armed conflict itself, these prerequisites are what make terrorism a viable and effective mode of operation in an asymmetrical confrontation.

As terrorism is perhaps the most asymmetrical form of political violence, it can be posited that these more specific prerequisites for the systematic and effective use of terrorist means in an armed con-

---

[34] On modernization as a 'traumatic experience' see Sztompka, P., *The Sociology of Social Change* (Blackwell: Oxford, 1993).

flict are directly related to the nature of the asymmetry between the main protagonists and especially to the characteristics of the armed non-state actors themselves. Even an explosive combination of extreme socio-political, economic and cultural imbalances with more tangible grievances (such as deep feelings of injustice, violations or lack of civil and political rights, or brutal government repression) does not necessarily provoke a non-state actor to confront the state by making politically motivated attacks against civilians. For that to happen, the state's opponent must be able to combine ideological determination with structural capability in a way that maximizes the group's comparative advantage if it chooses to resort to terrorism.

The degree of ideological commitment and indoctrination needed to 'justify' the use or threat of violence against civilians in a confrontation with a more powerful protagonist is significantly higher than for most other forms of violence widely practised by non-state actors. This high degree of indoctrination and the necessary 'justification' can only be provided by an extremist ideology. However, the fact that the ideological basis for terrorism may be provided by extremist ideologies of all types and origins—be it Maoism, anarchism, radical nationalism or Islamism—does not mean that any such ideology is inherently linked to terrorism or automatically produces it.

In the 19th century and throughout much of the 20th century terrorist means were most often used by adepts of various socio-revolutionary, anarchist and other radical left-wing ideologies. By the end of the 20th century, radical nationalism and religious extremism had emerged as the most influential ideological currents for groups employing terrorist means. At the global level, transnational terrorism is dominated by heavily politicized quasi-religious Islamist extremism. In many local and regional contexts, radical nationalism and religious extremism have merged, a combination that has sometimes been supported by local social norms and cultural traditions, such as the blood feud or remnants of slavery in clan-based societies.

In the modern complex, hybrid organizational structures of anti-system non-state actors (which increasingly display network features), the role of radical ideology as the glue holding together informally connected cells and elements has also become more important. To complement and reinforce ideological determination, a group that systematically uses terrorist means must also possess certain structural capabilities.[35] These capabilities extend beyond financial resources, technological skills and assets, access to arms and materials, the availability of trained professionals and so on. Rather, they refer to the specifics of the group's organizational model.

The structural development of many modern terrorist groups from the local to the transnational levels is marked by the spread of the features of network organizational forms. The more informal, flexible and fragmented is the organizational structure of such a group and the more network elements it incorporates, the greater are its comparative advantages in an asymmetrical confrontation with a state, with its more hierarchical structure. For some of the most advanced and novel organizational patterns, the network effect is being increasingly amplified by a unique phenomenon—the effective, multi-level coordination of multiple cells' activities exercised through generally formulated strategic guidelines. This phenomenon, which is not typical for either standard networks or classic hierarchies, is demonstrated by the multiple-cell, transnational post-al-Qaeda movement. These cells lack direct operational links but still manage to act and see themselves as parts of the same global movement.

In social sciences, the ideologies and the organizational forms of political violence have normally been addressed separately, by different schools of thought and within different theoretical frameworks. While the relationship between ideology and political violence has rarely been denied, purely instrumentalist, rationalist interpretations of ideology and of its role as a mere instrument in generating violence have become sidelined by the various schools of social constructivism, cultural anthropology and other disciplines that emphasize identity and beliefs. The focus on the role of 'agency' (which ranges from structure and organization to leadership and elites) in generating, stimulating and promoting violence and violent conflict mainly developed within the instrumentalist and rational choice tradition.[36] In contrast, this Research Report opts for a synthetic methodological approach that focuses on both ideological and structural aspects as the two most important, closely intertwined and mutually reinforcing characteristics of terrorism and terrorist actors.

---

[35] In this context, the term 'structural' refers to the way in which these groups are structured. Thus, this use is to be distinguished from the term's use as a synonym for 'fundamental', e.g. in 'structural causes'.

[36] For a more detailed discussion see chapters 2 and 3 (on ideology) and chapters 4 and 5 (on structures of terrorist organizations) in this volume.

The unique combination of extremist ideologies with certain structural capabilities and organizational patterns is the main precondition for the use of terrorist activities by militant non-state actors as a systematic tactic in asymmetrical confrontation. This combination is also their main comparative advantage vis-à-vis the principal protagonist. This precondition is more characteristic of terrorism than the broader structural and other fundamental causes of political violence in general.

A number of questions about the proper identification and categorization of the specific preconditions for terrorist activity could be asked in this context. One such question is whether resolving the violent conflict by addressing its main incompatibilities would automatically end the use of terrorist tactics. Finding a solution to the key issues and incompatibilities of the broader violent conflict is essential for undermining the foundations of terrorism as a tactic used in that conflict. However, even this may be insufficient to root out terrorism related to or generated by conflict. That will not happen unless the structural capabilities of militant groups employing terrorist means are fully disrupted and the role of extremist ideologies in driving their terrorist activities is effectively neutralized.

2. Ideological patterns of terrorism: radical nationalism

I. Introduction: the role of ideology in terrorism

In this Research Report, ideology is defined as a set of ideas, doctrines and beliefs that characterizes the thinking of an individual or group and may transform into political and social plans, actions or systems. While the ideological views and beliefs of those involved in terrorist activities are extremist by definition, this is probably the only aspect of the ideological basis and support of terrorism that is not disputed by analysts. All other issues related to the role of ideology for violent groups involved in terrorist activity remain unclear and are endlessly debated. There is no agreement even on the basic issue of whether there is any specific 'terrorism ideology' (i.e. whether terrorism itself is an ideology) or whether terrorists are, instead, driven by various extremist ideologies and exploit them to provide grounds for the use of terrorist means.

The idea of terrorism having its own, specific ideology is still relatively widespread in political and legal circles.[37] It also has its supporters in academia.[38] However, most scholars are sceptical about the idea. Their discussions on the subject are dominated by the alternative point of view that terrorism does not have a separate, specific ideology and is not itself an ideology in the way that socialism, fascism and anarchism are.

It must be kept in mind that the role of ideology in terrorism is a specific question which is part of the broader problem of the role of ideology in armed violence in general. When reviewing the concepts employed by anti-system actors to provide ideological grounds for the

[37] E.g. the Russian counterterrorism law defines terrorism as 'the *ideology* of violence and the practice of exerting pressure on decision making by state bodies, local government or international organizations, related to terrorizing the population and/or to other forms of violent action' (author's translation, emphasis added). Article 3 of Federal law of the Russian Federation of 6 March 2006 no. 35-FZ 'On countering terrorism', which entered into force on 10 Mar. 2006, published in *Rossiiskaya gazeta*, 10 Mar. 2006, <http://www.rg.ru/2006/03/10/borba-terrorizm.html> (in Russian).

[38] See e.g. Herman, E. S. and O'Sullivan, G., '"Terrorism" as ideology and cultural industry', ed. A. George, *Western State Terrorism* (Routledge: New York, 1991), pp. 39–75; and Soares, J., 'Terrorism as ideology in international relations', *Peace Review*, vol. 19, no. 1 (Jan. 2007), pp. 113–18.

use of violence in general and of terrorist means in particular it should be kept in mind that mere expressions of political support (e.g. for violence in the form of terrorism) do not amount to its ideological justification. Another basic starting point is that the use of violent, including terrorist, means by a certain group is not necessarily imposed on it by the nature of its final goals or by the principal ideology that it holds or claims to hold. The use of terrorist means by a group that considers itself to be, for example, Marxist does not mean that Marxism as an ideology calls for terrorism or should be associated with it in any way.

Terrorism is not an ideology; rather, it is a specific, hyper-extreme tactic of using or threatening violence. This tactic can be justified by terrorists within different ideological frameworks. Terrorists may sincerely believe in their guiding ideology, may be highly indoctrinated and may even be ready to sacrifice their own lives in a terrorist attack. However, in most cases they are not advanced or sophisticated ideologues. They may not even have a strong grasp of ideological nuances and may only vaguely understand the basic tenets of their extremist ideology. In other words, they are not so much people of the *word* as people of the *deed*.

The fact that terrorists do not have to be refined intellectuals—or, in the case of religious terrorists, advanced theologists—does not mean that terrorism is not ideologically driven. The definition of ideology used here goes beyond its narrow interpretation as 'abstract theorizing'. Ideology is not just a scripture or a set of theoretical pamphlets; it is a socio-political phenomenon associated with a socio-political context. It is not only a way of thinking that shapes a world view; it also provides the narrative and the means for translating individual and group grievances and experiences into socio-political action. Only the interconnection of ideological belief and politics in a particular political context explains how a radical ideology may serve as a basis for terrorist activity.

The evolving influences of ideologies

As terrorism in its various forms has evolved over time, the need to justify the use of terrorist means—and, consequently, the role of ideology as a provider of this justification—has grown. In the second half of the 19th century, political terrorism was still largely selective, with terrorists preferring to target specific individuals. Most commonly these were high-profile political or security figures, such as government ministers, or the 'tyrants'—kings and presidents—themselves. At that stage, terrorism was even partly justified by its advocates and perpetrators in 'humanitarian' terms: it was viewed as causing fewer innocent and accidental victims than, for instance, mass uprisings. Later, especially from the early 20th century, terrorism became less and less selective and eventually became a form of violence dominated by indiscriminate attacks on civilians. This made it even more pressing for terrorist groups and their leaders to provide ideological justification of their actions.[39]

In recent decades, the role of ideology for terrorist groups has also been growing due to their changing structural patterns, especially the rapid spread of network features. For complex network structures, the role of common ideological beliefs and goals as an organizing principle tends to be considerably more significant than for hierarchically structured entities. This common ideology acts as a structural glue that helps to connect often fragmented, informally linked elements and enables them to act as one movement.

Naturally, the ideologies that terrorist groups claim to use as a basis for their terrorist activities are related to their socio-political, nationalist or religious motivations, often employed in various combinations. However, regardless of a terrorist group's specific motivations and ideologies, their politico-ideological beliefs tend to display some common features. Among them is an idea that it is primarily the state which practises violence and terror. This argument has long been employed by terrorists of all sorts as a type of moral alibi. Another leitmotif common among terrorist groups can be summarized as 'the worse, the better'. In other words, the more disastrous and devastating the effects of terrorist attacks are and the more violent the reprisals from state authorities are, the better it is for the terrorists' cause. While all types of terrorist employ such arguments, the latter do not amount to a separate, specific ideology of terrorism.

---

[39] On the history of terrorism in the 19th and 20th centuries see e.g. Budnitsky, O. V., *Terrorizm v rossiiskom osvoboditel'nom dvizhenii: ideologiya, etika, psikhologiya (vtoraya polovina XIX–nachalo XX v.)* [Terrorism in the Russian liberation movement: ideology, ethics, psychology (the first half of the 19th–early 20th century)] (ROSSPEN: Moscow, 2000); and Laqueur, W., *A History of Terrorism* (Transaction: New Brunswick, N.J., 2001); and Hoffman (note 21).

In the 19th and much of the 20th centuries the ideologies of groups involved in terrorist activities were dominated by various radical socio-revolutionary, leftist and anarchist concepts. The ideologues of many left-wing terrorist groups, including socio-revolutionary organizations, often had eclectic views, integrating elements from different concepts and ideologies. These ranged from the anarchist motto of 'propaganda by deed', doctrines of the 19th century groups such as the Blanquists and revolutionary *narodniki*,[40] and radical Marxism, Stalinism, Trotskyism and Maoism to theories of anti-colonial struggle and the concepts of 'classic' rural or mountain and 'new' urban guerrilla activity. The ideologies of left-wing terrorists of the second half of the 20th century (such as the West German Red Army Faction and the Italian Red Brigades) did not include many motives and ideas beyond the 'classic' ideologies of radical revolutionary and anarchist groups of the 19th century. Among the few innovations were the Maoist concept of protracted civil war and, consequently, that of the use of terrorist means on a long-term, systematic basis rather than as a temporary tactic.

Over the 30-year period 1968–97,[41] communist/leftist groups were together responsible for the largest number of international terrorist incidents: 1869 in total.[42] In terms of the overall number of incidents, they were closely followed by nationalist/separatist groups, responsible for 1723 terrorist incidents. In contrast, the total of 497 terrorist

---

[40] Blanquism was a current in the 19th century revolutionary movement in France named after Louis-August Blanqui. He argued that a revolutionary movement can succeed without the broad armed support of the masses, primarily as a result of activity by conspiratorial groups of revolutionaries that resort to terrorism against authorities. In a broad sense, Blanquism may be a synonym for conspiratorial, rather than mass-based, revolutionary struggle. The *narodniki* were a socio-political movement in Russia in the 1870s–90s that advanced the concept of 'peasant socialism' and opposed tsarist autocracy. A small part of the movement, the organization Narodnaya volya (1879–84), prioritized political struggle and political violence and used terrorist means.

[41] For the period 1968–97, the MIPT Terrorism Knowledge Base provides data on international terrorist incidents only. See note 5. Some groups are categorized by the MIPT as both communist and nationalist (e.g. ETA) or as both religious and nationalist (e.g. Hamas or Lashkar e Toiba). The incidents, injuries and fataalites of such groups are thus included in the totals for 2 categories.

[42] The MIPT Terrorism Knowledge Base provides separate statistics for 'communist/ socialist' and 'leftist' groups, relying on DeticaDFI group taxonomy. See MIPT Terrorism Knowledge Base, 'TKB data methodologies', <http://www.tkb.org/DFI.jsp?page=method>. For the purposes of this study, data for these 2 types of group can be combined into 1 broad category—'communist/leftist'. The other categories of the Terrorism Knowledge Base used here are 'nationalist/separatist' and 'religious'.

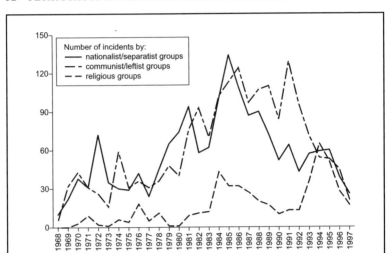

Figure 2.1. International terrorism incidents by communist/leftist, nationalist/separatist and religious groups, 1968–97

Source: MIPT Terrorism Knowledge Base, <http://www.tkb.org/>.

acts committed by religious groups was less than a third of that of either other category (see figure 2.1).

Despite secular leftist terrorism's responsibility for the highest number of international terrorist incidents at its second historical peak (from the 1960s to the 1980s),[43] the situation in terms of numbers of deaths is very different. Nationalist/separatist groups were responsible for the highest number of international terrorism-related deaths (3015) in the period 1968–97, almost twice as many as caused by religious terrorists (1640), while communist/leftist groups lagged far behind, with 829 deaths (see figure 2.2).

At the end of the 20th century some socio-revolutionary leftist groups whose ideology did not have a clear nationalist, let alone religious, aspect continued or started armed activity, including terrorism, especially in developing countries. Cases range from the Fuerzas Armadas Revolucionarias de Colombia (FARC, Revolutionary Armed Forces of Colombia) and Ejército de Liberación Nacional (ELN, National Liberation Army) in Colombia, who have been fighting con-

---

[43] The rise of socio-revolutionary and anarchist terrorism in the late 19th and early 20th centuries may be seen as the first historical peak of left-wing terrorism.

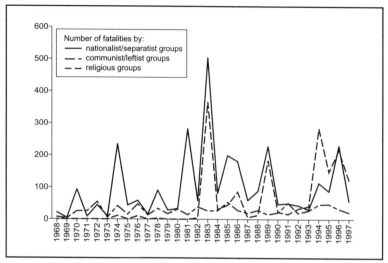

Figure 2.2. International terrorism fatalities caused by communist/leftist, nationalist/separatist and religious groups, 1968–97

Source: MIPT Terrorism Knowledge Base, <http://www.tkb.org/>.

tinuously for several decades, to the Communist Party of Nepal (Maoist) militants, who took up arms against the state in 1996. Throughout the 1990s some marginal leftist terrorist groups resurfaced in the developed world too, sporadically committing classic acts of 'peacetime' terrorism.

In the last decade of the 20th century, following the end of the cold war, communist, radical socialist and other leftist ideologies suffered an overall decline. This was mainly a result of the disintegration of the Soviet bloc, the end of the East–West ideological confrontation and the collapse of the bipolar world system. The role of these ideologies as a basis for groups involved in terrorist activity decreased. While communist and other leftist terrorism remained significant and even increased in 1998–2006 (see figure 2.3), its overall importance declined relative to the sharply rising nationalist and religious terrorism. This relative decline coincided in time and was connected with the gradual decline in state support for terrorism in line with the bipolar division. For much of the cold war period, many radical groups driven by communist and other leftist ideologies had enjoyed

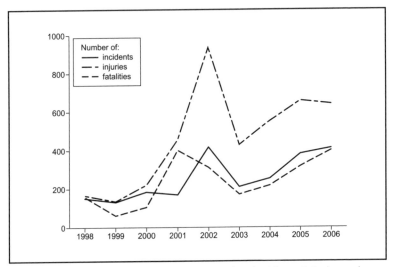

Figure 2.3. Domestic and international terrorism incidents, injuries and fatalities caused by communist/leftist groups, 1998–2006
Source: MIPT Terrorism Knowledge Base, <http://www.tkb.org/>.

some political and financial support from the states where those ideologies were dominant.[44]

In the 1990s, the ideological currents of radical leftism were increasingly replaced by radical nationalism, especially separatist ethno-nationalism, and by religious extremism, which became the two most influential ideological pillars of terrorism.[45] As noted above, even before the end of the 20th century the gap between nationalist and religious terrorism in terms of international terrorism fatalities was much narrower than in terms of incidents. In other words, even if religious terrorism resulted in much fewer international terrorist incidents, it appeared to have been more lethal than nationalist terrorism.

[44] In contrast, state support of religious and nationalist terrorism, e.g. in the Middle East and South West Asia, often continued. On state support of terrorism see e.g. Murphy, J. F., *State Support of International Terrorism: Legal, Political, and Economic Dimensions* (Westview: Boulder, Colo., 1989); and Byman, D., *Deadly Connections: States that Sponsor Terrorism* (Cambridge University Press: Cambridge, 2005).

[45] On ethno-nationalism and its distinction from civic nationalism see section II below.

In terms of international terrorism-related injuries, religious terrorism was even slightly ahead of nationalist/separatist terrorism.[46]

It can be tentatively argued that ideologies incorporating radical nationalism (including ethno-separatism) or religious extremism form a more favourable basis for inducing and 'justifying' the use of terrorist means than purely secular socio-political ideologies. In addition, an almost regular pattern can be observed: radical groups that systematically employed terrorism in an asymmetrical socio-political struggle that was not primarily driven by ethno-nationalist, national liberation or religious motives never succeeded in gaining and holding on to state power. This failure by, for example, Western anarchist and Russian socio-revolutionary terrorists is in contrast to the success of: (*a*) leftist and extreme right-wing opposition groups (revolutionary Marxists or social democrats in the late 19th and early 20th centuries and European fascists in the 1930s), which either did not employ terrorist means or did not use them systematically; and (*b*) nationalist, religious and ethno-religious groups that actively used terrorism as one of the main tactics in their armed struggle.

II. Radical nationalism from anti-colonial movements to the rise of ethno-separatism

The 19th and 20th centuries

As noted above, as terrorism emerged in the last third of the 19th century as a systematically employed asymmetrical tactic of political violence, it took no single form. Instead it was used by organizations of many political orientations in the name of many goals formulated in accordance with their various ideologies. Even at this early stage, terrorism was employed not only by socio-revolutionary groups, such as the Russian revolutionary *narodniki* or European and North American anarchists, but also by national liberation movements in the Balkans, India, Ireland and Poland.

In both the 19th and 20th centuries, most anti-colonial national liberation movements employed armed violence at some stage and in more than one form. Mahatma Gandhi's movement, which managed

---

[46] Religious groups were responsible for 10 863 injured victims in the period 1968–97 as opposed to 10 098 for nationalist/separatist groups. MIPT Terrorism Knowledge Base (note 4).

to achieve its goal of independence for India through non-violent means, was a rare exception. Broad national liberation movements often had extremist factions that, alongside other tactics, employed terrorist means, both against the colonizers and against the more moderate nationalists. In the mid-20th century, both prior to World War II and in the first post-war decades, terrorism was widely employed by anti-colonial and other national liberation movements in the Middle East, North Africa and parts of Asia. At that stage, several national liberation and nationalist groups that combined terrorist means with other violent tactics managed to achieve all or most of their declared goals. Some even came to power in their newly established states. The best known example of this period is the Algerian Front de libération nationale (FLN, National Liberation Front). The FLN led the armed struggle for independence from France after 1954 and at a certain point decided to turn to terrorist tactics in urban areas. It became the ruling party after Algeria's independence in 1962.[47]

Between 1968 and 1977—the first decade for which international terrorism statistics are available—the number of anti-colonial, other national liberation and ethno-separatist groups that used terrorist means in an international context (49 groups) was still slightly lower than the number of communist and other left-wing groups (58 groups).[48] However, nationalist terrorists were already responsible for 11 per cent more international terrorist incidents, 1.5 times more injuries and 2.2 times more fatalities than all communist and leftist groups.[49] The six deadliest nationalist groups of this period were all Palestinian organizations. The resort to terrorist means, including

[47] Terrorist means were also used by: the underground Jewish organization Irgun (Irgun Tseva'i Le'umi, National Military Organization, also known at Etzel), which fought for almost 2 decades for the creation of the state of Israel; and the Greek Cypriot insurgency movement Ethniki Organosis Kyprion Agoniston (EOKA, National Organization of Cypriot Fighters), which fought British rule in Cyprus in the mid-1950s and gained independence in 1960. Puerto Rican nationalist terrorist groups were also active in the years following World War II—launching terrorist attacks against US officials in the early 1950s and attempting to assassinate US President Harry S. Truman in 1950—but were not successful in advancing the goal of independence.

[48] Even if small anarchist groups are included, the total number of left-wing groups would not have exceeded 61. The number of terrorist groups with religious motives active in the same period did not exceed 5 and most of them (such as the Pattani United Liberation Organization in Thailand or the Moro National Liberation Front in the Philippines) combined religious and nationalist motivations and are also listed as nationalist/separatist groups. MIPT Terrorism Knowledge Base (note 4). On the MIPT definition of international terrorism see note 9.

[49] Calculation based on data from the MIPT Terrorism Knowledge Base (note 4).

international terrorism, by the Palestine Liberation Organization (PLO) and other Palestinian militant groups from the late 1960s until the 1980s demonstrated how to internationalize—and draw international attention to—a local asymmetrical armed struggle.

In sum, radical nationalism came to the fore alongside extreme left-wing ideologies as an ideology of groups that employed terrorist tactics. Even so, until the early 1980s various forms of radical left-wing internationalized socio-political ideology—ranging from Maoism to anarchism—still played a significant part as an ideological basis for groups engaged in terrorist activity. This was the case primarily in Europe, especially in France, West Germany, Greece and Italy, but also in other regions, from Latin America to Japan. In addition, many nationalist groups (e.g. the FLN in Algeria, the PLO and Euskadi Ta Askatasuna (ETA, Basque Homeland and Freedom) in Spain) effectively combined radical nationalism with leftist ideologies.[50] In the 19th and much of the 20th centuries such a combination was the rule rather than the exception. It was facilitated by the ambiguous approach to nationalism on the part of most socio-revolutionary ideologies, including Marxism. The only left-wing ideology that rejected nationalism was anarchism. Anarchists remained the most consistent and committed internationalists, proposing to replace nation states with cooperative communities based on free association and mutual assistance of people regardless of their ethnic and national origin.

Finally, radical nationalism, especially in its racist forms, was often an essential part of the ideologies of extreme right-wing socio-political organizations, including those that used terrorist means, such as the Ku Klux Klan movement in the USA.[51]

---

[50] E.g. ETA is categorized as both 'nationalist' and 'communist/socialist' in the MIPT Terrorism Knowledge Base (note 4).

[51] As noted in chapter 1 in this volume, section I, socio-political peacetime terrorism in general and right-wing terrorism in particular are not subjects of this study. As shown by the MIPT data for the period since 1998, right-wing terrorism has resulted in far fewer terrorist incidents, injuries and fatalities than nationalist, religious and left-wing terrorism. Right-wing terrorism is only mentioned here when combined with radical nationalism or religious extremism.

Into the 21st century

In the late 20th century this pattern changed. National liberation, especially anti-colonial, movements were replaced by radical ethno-nationalist movements, often with separatist aims. This new kind of ethno-nationalism was now rarely tied to left-wing ideology. Instead, it was more and more often linked to religious extremism. Along with the latter, radical ethno-nationalism and ethno-separatism moved to the fore as the ideologies most commonly employed by terrorist organizations. Ethno-separatist groups usually displayed a higher degree of intra-organizational coherence, continuity and resolve than, for instance, groups of a purely left-wing character. Ethno-separatist movements also proved able to remain active for decades without even changing their leaders.

In the early 21st century radical ethno-nationalism, and especially ethno-separatism, has retained its importance as one of the most widespread ideologies of groups employing terrorist means. However, it has gradually yielded primacy to religious, especially Islamist, extremism. Religious extremism has more and more often served as an ideological basis for terrorist groups active in more localized settings and, above all, for the emerging transnational violent Islamist movement. Sometimes, as in the case of the Islamic Movement of Uzbekistan (IMU), violent Islamism has served as a counterbalance and an alternative to nationalism; in other cases, as in Kashmir or Chechnya, it has been employed in combination with radical ethno-separatism.

Nationalism is a very powerful ideology that may provide the ideological framework for all kinds of ambitious political goals, including the break-up or formation of states.[52] It is also one of the most widespread ideologies in the world and takes many forms. These range from the more common passive forms to the more active ones that

[52] For some of the main interpretations of nationalism see: on modernization theory of nationalism—and nations—as a product of the emergence of the industrial society, Gellner, E., *Nations and Nationalism* (Blackwell: Oxford, 1981); on traditionalist explanations interpreting nation and nationalism as pre-existing (primordial) phenomena based on inherent cultural difference, Hobsbaum, E. and Ranger, T. (eds), *The Invention of Tradition* (Cambridge University Press: Cambridge, 1983); and on concepts that build on both traditionalist and modernist theories, but go beyond them (constructivist theories), Anderson, B., *Imagined Communities: Reflections on the Origin and Spread of Nationalism* (Verso: London, 1991) for a concept of nations as 'imagined political communities' and Smith, A. D., *Nationalism: Theory, Ideology, History* (Polity: Cambridge, 2001) on 'ethnosymbolism'.

imply political action in support of nationalist goals. These goals may range from cultural autonomy to separatism or irredentism. It is critically important to distinguish between such different forms of nationalism, and between its more moderate types and the more radical versions that may serve as an ideological basis for sustained political action. The latter may, under certain conditions, transform into the use of armed political violence.[53] Terrorism is just one—and not the most widespread—form of such violence.

This chapter mainly addresses ethnic (or ethno-political) nationalism as the most widespread—but not the only—type of nationalism advanced by armed non-state groups and movements. In contrast to civic nationalism, which views the nation as a voluntary and rational political association of citizens of a state bound by shared territory and institutions, ethno-nationalism emphasizes a common ethnic background as a basis for an organic nation.[54] According to ethno-nationalists, an ethnic group in a cultural and historical sense is identical to a nation as a political and state unit, and a common ethnic background is a necessary and sufficient basis for the formation of a separate state. The ultimate goal of ethno-nationalism is the creation of a separate state or quasi-state entity which is either mono-ethnic or in which the given ethnic group dominates.

In the post-colonial era, ethno-nationalism has largely replaced national liberation anti-colonial movements as the most evident and widespread version of radical nationalism. In multi-ethnic states, ethno-political movements have started to put forward more active demands that range from the redistribution of functions of governance and control over resources to the creation of separate states.[55] Excluding national liberation movements that fought against the colonial rule of European powers in the post-World War II period, over the period 1951–2005 a total of 79 ethno-nationalist movements representing

[53] On the relationships between nationalism and violence in general see Brubaker, R. and Laitin, D. D., 'Ethnic and nationalist violence', *Annual Review of Sociology*, vol. 24 (1998), pp. 423–52; and Beissinger, M., 'Violence', ed. A. J. Motyl, *Encyclopedia of Nationalism*, vol. 1, *Fundamental Themes* (Academic Press: San Diego, Calif., 2000), pp. 849–67.

[54] On civic and ethnic nationalism see Smith (note 52), pp. 39–42.

[55] See Tilly, C., 'National self-determination as a problem for all of us', *Daedalus*, vol. 122, no. 3 (summer 1993), pp. 29–36; Simpson, G. J., 'The diffusion of sovereignty: self-determination in the post-colonial age', *Stanford Journal of International Law*, vol. 32 (1996), pp. 255–86; De Vries, H. and Weber, S. (eds), *Violence, Identity, and Self-Determination* (Stanford University Press: Palo Alto, Calif., 1997); and Moore, M. (ed.), *National Self-Determination and Secession* (Oxford University Press: Oxford, 1998).

territorially concentrated ethnic groups were engaged in armed strug-gle for autonomy or independence from central governments.[56] At the end of 2006, such 'self-determination' movements were engaged in 26 active armed conflicts.[57] Among other tactics, these movements—in Chechnya, Kashmir, Mindanao or Sri Lanka—have increasingly started to use terrorist means to achieve their goals.

This does not mean that civic nationalism cannot be radicalized to the point where it turns to violence or even terrorism. On the contrary, civic nationalism in its radical forms, particularly on the part of the state, has a long history of deadly and mass violence both against other states and against ethnic minorities. Among non-state actors, the notion of nascent civic nationalism was more appropriate than ethno-nationalism to largely secularized anti-colonial movements, including those that employed terrorist means.

In the post-colonial era, in addition to narrowly ethno-nationalist and ethno-separatist movements, another form of armed nationalism has been national liberation from foreign occupation. While some such movements may be dominated by the prevailing ethnic group in the 'occupied' state, in contrast to radical ethno-nationalists they are usually multi-ethnic (and inter-confessional). However, the supra-ethnic nature of most modern armed national liberation movements, especially in Muslim-populated regions, is not civic in nature and is increasingly tied to their Islamicized character. Ongoing armed national liberation movements have either continued from the 20th century after having undergone some changes (e.g. Islamicization in the case of the Palestinian armed resistance) or have newly emerged in the early 21st century (such as the post-2003 resistance in Iraq).

[56] Hewitt, Wilkenfeld and Gurr (note 2), p. 33. The CIDCM data sets on peace and conflict issues are the primary sources of data used in this chapter.

[57] Hewitt, Wilkenfeld and Gurr (note 2), p. 33. 'Self-determination movement' is the term used to denote ethno-nationalist movements in reports by the Center for International Development and Conflict Management (CIDCM). See also the previous CIDCM report Marshall, M. G. and Gurr, T. R., Center for International Development and Conflict Manage-ment (CIDCM), *Peace and Conflict 2005: A Global Survey of Armed Conflicts, Self-Determination Movements, and Democracy* (CIDCM: College Park, Md., 2005), <http://www.cidcm.umd.edu/publications/publication.asp?pubType=paper&id=15>.

III. The 'banality' of ethno-political conflict and the 'non-banality' of terrorism

Any analysis of the ideological basis of ethno-nationalist terrorism must focus on the most radical separatist forms of ethno-nationalism that involve—and require—the sharp polarization of ethnic identities. However, ethno-political extremism per se does not necessarily imply or require the use of organized armed violence. The role of ideology in the processes of radicalization of an ethno-nationalist movement, and its resort to armed violence in general and terrorism in particular, needs to be further clarified.

Exploring the process of mobilization of violence and identifying the point where, for example, ethnic polarization, inter-ethnic tensions and hostility turn into armed violence is one of the most complicated analytical problems in conflict studies. It is also one that remains largely unresolved. Any analytical calculation that includes nationalism, an 'ethnic factor' and associated violence requires a great deal of caution. This is particularly so when trying to generalize about nationalist violence, with its multiplicity of forms and manifestations. These range from genocides, riots and inter-communal crowd violence to acts of terrorism, which is far from being the most common and mass-based form. It also needs to be recalled that, unlike some other types of nationalist violence, such as genocide, terrorism as it is defined here can only be carried out by non-state actors.

Despite frequent references to ethno-political terrorism in political and public discourse, there is surprisingly little research on the phenomenon. Most serious work on terrorism as a tactic of violent ethno-nationalist movements has taken the form of specific case studies.[58] Attempts to conceptualize this form of violence have been few and superficial.[59] This gap in research can only partly be explained by the lack of attention paid by many political scientists and experts in conflict studies to the specifics of terrorism as compared to other forms of violence. It is also a good illustration of the more generic problem of explaining the relationship between nationalism and vio-

---

[58] In the European context see e.g. Reinares, F., *Patriotas de la Muerte: Quiénes han militado en ETA y por qué* [Patriots of death: who joined ETA and why] (Taurus: Madrid, 2001); and Alonso, R., *The IRA and Armed Struggle* (Routledge: London, 2006).

[59] See e.g. Byman, D., 'The logic of ethnic terrorism', *Studies in Conflict and Terrorism*, vol. 21, no. 2 (Apr.–June 1998), pp. 149–70.

lence and identifying specific mechanisms for the mobilization of nationalist violence.

Naturally, in the Western literature most attention has been paid to ethno-political terrorism of West European origin, such as has been practised for decades by ETA and the Irish Republican Army (IRA) and its many offshoots. In both cases, an ethno-nationalist motivation—aggravated by irredentist and confessional motives in the case of the IRA—has prevailed over other, socio-political, motivations and goals, such as the leftist and anti-fascist motives that formed an integral part of ETA's ideology. However, most explanations of the phenomenon of ethno-political terrorism by these and some other Western groups have hinged on the extent of ethnic, or ethno-confessional, polarization and the ways in which it was exploited for political purposes. The most that appears to have been concluded regarding the nature of the link between ethnic factors and terrorism is that, the sharper the societal divide along ethnic lines is, the more fierce and bitter is the resulting armed ethno-political confrontation and the more it is likely to take the form of terrorism.

These explanations are hardly sufficient: the analytical problem being addressed here cannot be solved by references to the particular brutality of ethno-political conflicts alone or the allegedly more aggressive nature of ethno-nationalism compared to other radical ideologies. Extreme ethno-nationalism may, indeed, be seen as a more powerful and sustainable radical ideology than some socio-political 'internationalist' left-wing currents, especially in the late 20th and early 21st centuries. However, the 'superior' nature of the mobilizing and persuasive powers of radical nationalism is less evident if it is compared to religious extremism, especially at the transnational level.

Terrorism is not, in fact, a natural outgrowth or a necessary attribute of the extreme bitterness of an ethno-political conflict. Terrorist means were rarely used during the conflicts in the Balkans during the 1990s and have not been employed in cases of genocide in the Great Lakes region of Africa. Rather, terrorist means have been systematically used by ethno-separatist movements engaged in protracted, chronic conflicts. This has been the case in Chechnya, Kashmir, Sri Lanka and elsewhere. As noted above, the only other type of modern conflict where nationalist terrorism is employed systematically is the armed national liberation struggles (e.g. by groups of the Palestinian and Iraqi resistances).

The role of the ethnic factor in armed violence

As noted above, the analysis of different forms of nationalist violence, including terrorism, remains one of the least explored areas in the study of nationalism and violence. It seems that the best way to explain the specifics of nationalist terrorism is through comparison with—and contrast to—other, more widespread forms of nationalist violence.

Since the closing years of the cold war there has been a generally stronger emphasis on ethno-nationalism and ethnic factors as drivers of contemporary armed conflicts.[60] With the proliferation of ethno-political conflicts in the early post-cold war period, the ethnic factor came to be viewed as a force that inherently directs an ethnic group towards aggression against other ethnic groups. This approach is rooted in the much criticized and relatively marginalized primordialist 'cultural difference' school of the 1970s and 1980s, which argued that nationalist violence is inherent in—and a natural progression of—cultural difference.[61] However, the unique role of an ethnic factor and ethno-nationalist ideology in causing armed violence, including terrorism, should not be overemphasized for a number of reasons.

First, the combination of ethno-nationalism and *non-violence* appears far more common than that of ethno-nationalism and violence. The best efforts of researchers to compare the numbers of real (i.e. active) and potential inter-ethnic and inter-communal conflicts show that most ethnic groups manage to live in peace with one another, despite frequent tensions between them. For instance, studies by James Fearon and David Laitin based on evidence from Africa since 1979 show that only 0.28 per cent of real and potential inter-ethnic tensions resulted in armed conflict.[62]

The available data show that most nationalist conflicts do not result in large-scale violence. In most conflicts, only a part—usually small—of a nation or an ethnic group partakes in violence. An even

---

[60] The advocates of this approach range from scholars such as Donald Horowitz and Michael Ignatieff to publicists such as Robert Kaplan. See Horowitz, D. L., *Ethnic Groups in Conflict* (University of California Press: Berkeley, Calif., 1985); Ignatieff, M., *Blood and Belonging: Journeys into the New Nationalism* (Farrar, Straus and Giroux: New York, 1993); and Kaplan, R. D., *The Ends of the Earth: A Journey to the Frontiers of Anarchy* (Random House: New York, 1996).

[61] On primordialism see also note 52.

[62] Fearon, J. D. and Laitin, D. D., 'Explaining interethnic cooperation', *American Political Science Review*, vol. 90, no. 4 (Dec. 1996), pp. 715–35.

smaller proportion of ethno-nationalist movements, including ethno-separatist or 'self-determination' movements, chooses to resort to armed violence. According to data from the University of Maryland Center for International Development and Conflict Management (CIDCM), by 2005 only 25 such movements were involved in armed conflicts. In 54 other cases organizations that claimed to represent territorially concentrated ethnic groups tried to acquire a greater level of autonomy or self-determination for their groups through peaceful political means. Another 23 movements combined non-violent means—such as building a mass support base, identifying and publicly defending group interests, taking part in election campaigns or launching peaceful protest actions—with sporadic, isolated acts of violence that fell short of armed confrontation. Most such movements were active in democratic Western countries (e.g. Flemings and Walloons in Belgium and Catalans in Spain). However, some were ethno-nationalist movements using or advocating sporadic acts of violence against more rigid and authoritarian regimes. Examples of the latter include Mongols, Tibetans and Uighurs in China and Pashtuns and Sindhis in Pakistan.[63]

Even if ethno-political movements resort to violence, terrorism is not the most common and widespread form of such violence. It also usually enjoys less public support than, for instance, rebel attacks against government military and security targets. While an ethno-political insurgency movement as a whole may enjoy broad support among its ethnic base, those radical parts that systematically employ terrorist means usually do not have the same level of support.

Second, findings about ethno-political groups in conflict hint at the extremely complicated nature and multiple causes of those conflicts that are commonly—and often simplistically—identified as 'ethno-political'. Such conflicts usually result from a combination of inter-related socio-political, economic and cultural factors, issues of identity, and so on. Ethno-nationalism is not necessarily the only, or even the most important, driver.[64]

---

[63] Marshall and Gurr (note 57), pp. 21–22, 25, 27.

[64] See e.g. Hardin, R., *One for All: The Logic of Group Conflict* (Princeton University Press: Princeton, N.J., 1995); Reno, W., *Warlord Politics and African States* (Lynne Rienner: Boulder, Colo., 1998); Mueller, J., 'The banality of "ethnic war"', *International Security*, vol. 25, no. 1 (summer 2000), pp. 42–70; and Fearon, J. D. and Laitin, D. D., 'Ethnicity, insurgency and civil war', *American Political Science Review*, vol. 97, no. 1 (Feb. 2003), pp. 75–90.

The notion of 'purely ethnic' violence is therefore something of an abstraction. If an ethno-nationalist group is engaged in armed struggle, that does not necessarily mean that the violent conflict has no other causes, motivations and participants. So-called ethnic violence is often an integral part of the broader, complex mix of different forms of political and profit-driven, organized and unorganized, direct and structural violence. This phenomenon has been best captured by John Mueller's theory of the relative *banality*—that is, the unexceptional nature—of armed conflicts with an ethno-political form.[65] The idea of the banality of ethnic conflict effectively challenges the thesis of ancient ethnic hostility as the driving force of armed conflicts—a primordialist explanation that resurfaced in the post-cold war era.[66] Even long historical experience of confrontation, reinforced by systematic propaganda by ethno-political elites and leaders, does not guarantee support for violent ethno-nationalism, especially in the form of terrorism, by the broader population. This may be true even at the peak of the fiercest of armed conflicts, especially if explicit discrimination on an ethnic basis was not the main, direct motivational cause of that conflict.

Another argument in favour of the thesis of the banality of ethnic violence is that in complex multi-causal and multi-level armed confrontations it is often intimately intertwined with other forms of violence. For instance, the widespread combination of ethnic strife with criminal violence has become typical for many conflict and post-conflict areas. It may develop to a point where acts of violence driven by ethnic hatred are often mistaken for or cannot be distinguished from violent crimes committed by people of one ethnic group against those of another primarily for material gain, as often happened in the Balkans in the 1990s and early 2000s.

Finally, the very idea that armed violence in an ethno-political form is a kind of aberration or a radical deviation from a presumed norm of peace raises some questions. The rise of ethno-separatism in the so-called Third World in the post-World War II period, especially in post-colonial Africa and Asia, should have surprised no one. The vast

---

[65] Mueller (note 64). This banality of ethno-political violence should not be confused with or be reduced to 'rationality', i.e. the instrumentalist, rational-choice interpretation of such violence as nothing more than a rationally employed instrument to achieve a group's goals.

[66] On the primordialist explanation see e.g. Hobsbaum, E., *Nations and Nationalism since 1780: Programme, Myth, Reality* (Cambridge University Press: Cambridge, 1990). See also note 52.

majority of the new states were artificial entities with borders arbitrarily drawn by their former European colonial rulers. Despite this, the conceptual thinking on the subject, undertaken mostly in the West, was for a long time dominated by the thesis of ethno-political violence as something exceptional and specific to these local contexts. The popularity of this thesis may be partly explained by perceptions of the relatively atypical nature of large-scale ethno-nationalist violence for many developed states themselves, as compared to many less developed and more ethnically diverse countries affected by ethno-political conflicts.[67] The perception that large-scale ethno-nationalist violence is mostly restricted to underdeveloped regions may not be very accurate, but it has some basis in the much lower levels of ethno-political violence in Western countries.

Terrorism as extreme violence within violent extremism

While radical ethno-nationalism, and especially ethno-separatism, may well serve as an ideology for groups that employ violent means to achieve their political goals, it does not necessarily lead to violence. Even if it does, not all of the violent ethno-political groups in regions such as Central and East Africa, Central, South and South East Asia, and Eastern Europe necessarily use terrorism. In addition, so-called ethnic violence is most often a mixture of many socio-political, economic, cultural and identity factors and influences.

Given that ethno-nationalist movements are not intrinsically violent, the question arises of why some of these movements resort to terrorism. The need to answer this question leads back to the thesis of the banality of ethno-political violence, especially outside the Western world. In contrast to most Western states, for many of the mostly multi-ethnic states in other regions of the world ethno-political violence is not seen as an exceptional phenomenon but rather as just one of the common, chronic and recurring manifestations of broader pattern of protracted, complex violence.

Against this background, the key to understanding why some radical ethno-nationalists resort to terrorism may be summarized as follows. If there are grounds to assert the relative banality of ethno-political violence, then the main characteristic of terrorism is precisely its *non-banality*, even within the broader cycle of violence. In order to

---

[67] See e.g. Horowitz (note 60).

play its role for the violent actors, terrorism must be perceived as excessive, an aberration. A terrorist act must be a spectacular event that goes far beyond routine practices, politics, behavioural patterns and even routine violence.

Nonetheless, the dividing line between the banality of ethnic violence and non-banality of terrorism may be very thin. Despite this, the main distinguishing feature and comparative advantage of terrorist tactics is precisely its extraordinary nature. It is event-centred in the sense that it aims at causing a spectacular and shocking political event whose impact goes far beyond its direct human and material damage. This non-banality manifests itself through a set of characteristics of terrorist tactics. They include implacable ruthlessness, a readiness to mount indiscriminate attacks and target innocent civilians, often in large numbers, the unconventional use of conventional means, and the demonstrative and communicative nature of terrorist acts. The impression that terrorism is non-banal should be strong enough to contrast with other forms of violence that are more common, widespread and mass-based and could be perceived as more acceptable. Whenever terrorism becomes customary and banal, it loses much of its political effect. Terrorism is 'abnormal' violence; it makes sense for the perpetrators inasmuch as it can be perceived as *extreme violence within violent extremism*.

In the situation of relative civil peace, general state functionality and more or less effective accommodation of ethnic minorities that is inherent to most developed Western countries, terrorism as a tactic of violent ethno-political movements is perceived as an aberration by default. This partly explains why ethno-nationalists in those few Western state that still face militant separatism often choose terrorism over other forms of violence.[68] However, in the other parts of the world, where the bulk of global terrorist activity occurs, violence is chronic, institutionalized and often perceived as a norm. This is particularly the case in conflict and post-conflict areas in developing, underdeveloped, weak, failed and dysfunctional states. In some areas

---

[68] Of the 221 nationalist/separatist groups that used terrorist means in 1998–2006, 37 were groups in Western countries, mostly tied to 3 separatist causes—in Corsica, Basque-populated regions of Spain and France, and Northern Ireland). Twelve of these groups were responsible for terrorist attacks that caused deaths, usually within the range of 1 to 3 fatalities. The 2 groups that have killed a larger number of civilians in terrorist attacks were ETA (responsible for 54 fatalities) and the Real IRA (30 fatalities). Calculations are based on the MIPT Terrorism Knowledge Database (note 4).

affected by protracted armed conflict, the more general distinction between normal and abnormal social and political behaviour, both violent and non-violent, becomes blurred and what was perceived as normal may become distorted beyond recognition.

In these areas, the use of terrorist means cannot, by definition, guarantee the same impression of non-banality and excess. Depending on the ruthlessness of a particular armed conflict, terrorism may not necessarily be perceived as extraordinary or extreme violence. It may, in fact, be outmatched in terms of cruelty, deadliness, the number of people affected and even the broader public effect by other forms of ethno-political violence such as mass ethnic cleansings or genocide. The boundaries between asymmetrical terrorism and symmetrical violence, such as inter-communal and sectarian violence, may also become increasingly blurred.[69] In these parts of the world, terrorism has a better chance of retaining its non-banal nature where there is a sharp contrast between the two co-located but radically different socio-political and cultural systems or communities—for instance, more modernized (and Westernized) and more traditional communities.[70]

The conclusion that follows is that nationalist terrorism in general, and ethno-nationalist terrorism in particular, is likely to be more effectively employed wherever it retains the effect of non-banal violence, going beyond the limits normally applied to the more customary and banal forms of violence.

IV. Real grievances, unrealistic goals: bridging the gap

The above explanation of violent ethno-nationalists' resort to terrorist means is not the only one and it is not the one most directly related to ethno-nationalist ideology as such. It should be supplemented with a second explanation. This starts from an assumption that the prospects for the final and complete achievement of radical ethno-nationalist goals—ultimately centred on separatism and the creation of a new state—are limited. In the modern world, most ethnic groups have not

---

[69] This has happened e.g. in post-2003 Iraq, where terrorism has become increasingly intertwined with, and indistinguishable from, inter-communal sectarian strife. See also chapter 3 in this volume, section V.

[70] Examples include the divide between the francophone parts of Algeria, which were colonized by citizens of metropolitan France, and the rest of the country; and that between the Israelis and Palestinians.

formed their own states. A situation in which every ethnic group would be entitled to a separate state is simply inconceivable. Despite the relatively high mobilization potential of ethno-separatism, the formation of an independent state on the basis of a separatist movement has generally remained the exception, rather than the rule in the post-colonial era. Any such case is handled separately, and at length, by the international community and is rightly viewed as a potentially destabilizing precedent. Against this background, even if radical ethno-nationalism is backed by sustained armed violence, it does not guarantee the formation of a separate, mono-ethnic state.

According to the CIDCM data, of the 71 self-determination conflicts in 1951–2005, ethno-separatist movements managed to gain an internationally recognized independent state as a result of armed violence in only five cases: the Bengalis in Pakistan (Bangladesh was formed in 1971), the Slovenes and the Croats, whose successions from Yugoslavia were recognized in 1991–92; the Eritreans in Ethiopia in 1993; and the East Timorese in Indonesia (Timor-Leste became independent in 2002).[71] Several quasi-state entities that were formed by separatist and irredentist movements and enjoy *de facto*, but not internationally recognized, independence could be added to this list. These include Abkhazia and South Ossetia (in Georgia), Kosovo (in Serbia), Nagorno-Karabakh (in Azerbaijan), Somaliland (in Somalia), Trans-Dniester (in Moldova) and the Turkish Republic of Northern Cyprus. It should also be noted that in some of these cases—in Somaliland and Trans-Dniester—separatism was primarily driven by factors of socio-politics, economics and historical regionalism, rather than of ethnicity.

In most other cases the most that a radical ethno-nationalist group with separatist aims may realistically hope to achieve is some form of redistribution of power within a state. The resulting settlement often takes the form of a federal power-sharing arrangement or regional autonomy. While no multi-ethnic state can guarantee the absolute equality of all ethnic groups, more equitable federal arrangements are increasingly widespread, not only in the developed world but also in developing countries. They can provide for the peaceful coexistence of various groups and deprive extremists of opportunities to mobilize violence on an ethnic basis. These frameworks allow ethno-nationalist movements, including those that once took up arms to fight for their

---

[71] Marshall and Gurr (note 57), p. 23–24.

cause, better access to the central government's decision-making processes and the chance to gain greater regional autonomy. In sum, despite a number of highly publicized post-cold war cases of sustained ethno-separatism (such as in Kosovo and Abkhazia), ethno-nationalist movements with separatist goals rarely achieve a revision of internationally recognized borders.[72]

Research shows that terrorism is most closely connected to political factors and conditions such as chronic discrimination, including discrimination on an ethnic basis or the violations or absence of civil and political rights.[73] While the ultimate political goals and motivations of radical ethno-nationalist movements are at least to some extent based on these and other real grievances, this does not mean that these goals are realistic. If, for instance, the goal is to achieve a broader and more equitable representation in state structures or a greater degree of autonomy for an ethnic group, then that goal is usually achievable in some way. It may even have relatively high chances of being realized, whether through the normal political process or an armed struggle. If, however, the goal is the creation of an independent state, then in most cases its chances of being achieved are much lower, regardless of the methods that are used to advance it. It is thus unsurprising that many ethno-separatist conflicts are protracted confrontations that can last for decades without any realistic prospects of the separatists achieving their ultimate goal of independent statehood. The average duration of the 25 such conflicts that were active in the early 21st century was 27 years.[74] Even though the number of ethno-separatist conflicts has declined since the early 1990s,[75] few can be seen as finally resolved.

On the one hand, real grievances such as foreign occupation or repressive actions by the state or dominant ethnic group create the necessary conditions for mobilization of ethno-political violence. They can be effectively seized on by ethno-nationalists leaders and ideologues. On the other hand, this strong mobilization potential collides with the inherently low chances of ethno-nationalists' ultimate

---

[72] This pattern is confirmed by the CIDCM data. See e.g. Hewitt, Wilkenfeld and Gurr (note 2), p. 38.

[73] See Lia, B. and Skjølberg, K., *Causes of Terrorism: An Expanded and Updated Review of the Literature* (Norwegian Defence Research Establishment: Kjeller, 2005), <http://rapporter.ffi.no/rapporter/2004/04307.pdf>.

[74] Marshall and Gurr (note 57), pp. 26–27.

[75] Marshall and Gurr (note 57). This trend was already clear by the end of the 1990s. See also Gurr, T. R., 'Ethnic warfare on the wane', *Foreign Affairs*, vol. 79, no. 3 (May/June 2000), pp. 52–64.

stated goal—independence—being achieved, even through the use of violent means. This collision is a recipe for further radicalization of violence, at least by the extremes of the ethno-nationalist movement, and explains the need for the violence to assume increasingly asymmetrical forms such as terrorism. In other words, the more realistic are the ethno-nationalist movement's political goals, the less is the need for it to resort to terrorist means and the lower are the chances that terrorism will become one of the main tactics employed by ethno-nationalists.

Of critical importance is how realistically ethno-separatists perceive their final goals, regardless of which specific factors appear to them to make the achievement of their ultimate goals more or less realistic. International support is one of several factors that may affect these groups' proclivity to employ terrorist means. Two examples illustrate the diametrically opposite influences that this factor may have on separatists' perceptions of their chances of gaining international recognition of an new independent state.

An unusual characteristic of the situation in Kosovo from the late 1990s was the high level of direct international support for armed separatists, primarily from the USA and some other leading Western states, as well as its partners in the North Atlantic Treaty Organization (NATO). Naturally, such high levels of external support increased the likelihood of the Albanian Kosovar ethno-separatists' achieving their goal of independence, at least in their eyes. It should then not be surprising that, despite the many forms that armed violence has taken in Kosovo (guerrilla tactics, ethnic cleansing and inter-communal warfare), the armed ethno-separatist movement saw no need to resort to the 'extraordinary' tactics of terrorism.

That same factor of external support can also play an opposite role, even in cases where nationalist movements have a broader national liberation, rather than narrowly ethno-separatist, character. There is broad international recognition of the right of the Palestinian people to a sovereign state that is to include some of the territories still occupied by Israel.[76] Despite this, the continuing resistance to the Israeli occupation of Palestinian territories, which involves the systematic use of terrorist means, has little chance of achieving that goal—at least as

---

[76] This right had been repeatedly confirmed by United Nations Security Council resolutions. E.g. UN Security Council Resolution 242, 22 Nov. 1967; and UN Security Council Resolution 338, 22 Oct. 1973.

long as Israel enjoys the support of the USA. In a situation such as this, where there is a wide gap between a high nationalist mobilization potential among the Palestinians and a low chance of the nationalists' ultimate goal being realized, the systematic employment of terrorist means is not surprising. In the Palestinian case, the use of terrorism means by the more radical parts of the nationalist resistance, both the secular and the Islamist, is likely to continue as long as this gap persists.

V. Conclusions

One of the main prerequisites for a radical section of an ethno-nationalist movement to resort to terrorism is a significant gap between, on the one hand, the objective chances of achieving its final goal of an independent state and, on the other, its own unrealistic perception of how likely that goal is to be met. The main task of the extremist ethno-nationalist ideology is to 'virtually' bridge that gap.

However, it would be an exaggeration to attribute ethno-nationalist terrorism mainly to the effects of systematic radical nationalist propaganda. Such an oversimplification overlooks the roles of real sociopolitical grievances as the most direct causes of political discontent that takes ethno-nationalist form and of real (or perceived) threats to the well-being, identity or even survival of a certain ethnic group. The most critical role of radical ethno-nationalist ideology is to avert or disguise this fundamental collision between real grievances that cause conflict of an ethno-political form and the final, and probably unachievable, goals of the radical ethno-separatists.

In this way, ideological extremism in the form of radical nationalism provides the mechanism for the gradual further radicalization of the movement and its resort to terrorist tactics. The use of terrorist means can be particularly effective if, compared to other forms of violence in the context of the same ethno-political conflict, terrorism is perceived as a non-banal, 'extreme', 'abnormal' violence.

As noted above, the problem does not boil down to any single, specific ideology. Rather, a set of certain ideological postulates and characteristics is required, some of which may be easily formulated and defended within the framework of radical ethno-nationalist discourse. These characteristics include the infamous postulate that 'the worse, the better'; the tendency to encourage destructive self-expression; and the tendency to blame the state for all forms of violence and to view it as the source of all 'evils'. While these characteristics can often be separately identified in the ideologies of radical organizations that do not use terrorist means, their combination is usually an ideological recipe for terrorism.

Finally, even if the real underlying causes for ethno-nationalist discontent are effectively removed and the state's policy or a peace process accommodates most of the ethno-nationalists' demands, it does not guarantee an end of terrorist activity by the most radical ethno-separatists. These policies also cannot prevent the emergence of splinter groups that may continue to use terrorist means. However, ideologically the state has something in common with even the most radical and violent ethno-separatists, including those that employ terrorist means—the central focus on the state itself as the main point of reference. On the one hand, this turns violent ethno-nationalists into some of the worst enemies of many existing states and societies, especially the multiethnic ones. On the other hand, it also makes radical ethno-nationalism a recognizable enemy for the state. This enemy exists in the same dimension, or framework, as the state itself: violent ethno-nationalists see themselves as part of the same state-based world which they eventually aspire to join on equal terms. They operate within the same discourse as the state itself and accept and even glorify the very notion of the nation state. All that ethno-nationalists ultimately aspire is to form a state that could be on equal status terms with other states. This is in sharp contrast to the transnational versions of some of the radical religious and quasi-religious ideologies discussed in the next chapter.

3. Ideological patterns of terrorism: religious and quasi-religious extremism

I. Introduction

In the 1990s, following the collapse of the Soviet bloc, the end of the cold war and the decline of leftist movements, a global vacuum in secular protest ideology emerged. This vacuum quickly started to be filled with radical currents—the explicitly extremist ethno-nationalist or religious ideologies.

Much has been written about the 'sharp' rise of 'religious terrorism' during the last decades of the 20th century and about its growing internationalization and international impact. However, to back this thesis most analysts choose not to look at the available data directly. The same few pieces of quantitative evidence are usually quoted, covering the same period of time (from the late 1960s until the mid-1990s) and derived from the same sources—most commonly from terrorism experts Bruce Hoffman and Magnus Ranstorp. For example, these experts' reference to the fact that over the 30-year period until the mid-1990s the number of radical fundamentalist religious groups professing various confessions tripled has been reproduced in a number of analyses. These analyses also note that there was an increase in terrorist groups of an 'explicitly religious' character from virtually no such groups in 1968 to a quarter of all terrorist organizations by the early 1990s (somewhat declining to 20 per cent of approximately 50 active terrorist groups in the mid-1990s).[77]

Nevertheless, the number of groups using terrorist means is just one of several indicators of terrorist activity. It is not the most important, is one of the most ambiguous and should only be considered in con-

---

[77] E.g. Hoffman, B., ' "Holy terror": the implications of terrorism motivated by a religious imperative', *Studies in Conflict and Terrorism*, vol. 18, no. 4 (Oct.–Dec. 1995), p. 272—for an earlier version see Hoffman, B., *'Holy Terror': The Implications of Terrorism Motivated by a Religious Imperative* (RAND: Santa Monica, Calif., 1993), <http://www.rand.org/pubs/papers/P7834/>, p. 2; Ranstorp, M., 'Terrorism in the name of religion', *Journal of International Affairs*, vol. 50, no. 1 (summer 1996), pp. 41–62; Hoffman, B., 'Terrorism trends and prospects', I. O. Lesser et al., *Countering the New Terrorism* (RAND: Santa Monica, Calif., 1999), <http://www.rand.org/pubs/monograph_reports/MR989/>, pp. 16–17; and Hoffman, B., 'Old madness, new methods: revival of religious terrorism begs for broader U.S. policy', *RAND Review*, vol. 22, no. 2 (winter 1998/99), pp. 12–17.

junction with the other main indicators of terrorism and in a specific national and political context. The number (or size) of groups may not be in direct relation to the overall level of terrorist activity of a certain type in a certain context. The nature, level of organization, ideological consolidation, militant proficiency, and public relations and propaganda sophistication of these groups may be of greater importance for the effective and systematic use of terrorist violence.[78] As far as terrorism in concerned, it is not quantitative indicators alone that matter, but it is still worth considering a combination of all available indicators, especially as the overall picture obtained from analysing these data is somewhat more complex and nuanced.

For example, over the period of almost four decades (1968–2006) for which continuous data on international terrorist incidents are available at the time of writing, there were only four years in which the number of such incidents carried out by religious groups worldwide exceeded those committed by nationalist/separatist groups. What is most worrying is that three of these four years are the most recent ones, 2004–2006, with 1994 as the fourth (see figure 3.1). In terms of international terrorism-related casualties, it was only in 1993 that religious terrorism first accounted for more fatalities than nationalist terrorism. This pattern has continued since 1993 with the exception of only two years—1996 and 1999—when nationalist terrorism resulted in higher numbers of deaths (see figure 3.2).

In contrast, at the domestic level, over the period 1998–2006 nationalist/separatist terrorism resulted in a significantly higher number of attacks than religious terrorism: 2808 as opposed to 1824, or 54 per cent more.[79] This is unsurprising, given the primarily domestic focus of nationalism. However, even at the domestic level, religious terrorism was somewhat more deadly than nationalist/separatist terrorism. Nationalist/separatist groups accounted for almost the same total number of injuries—12 812 as opposed to 12 863 by religious groups—but for a lower number of fatalities—5648 as opposed to 6607 by religious groups (or 15 per cent fewer).

---

[78] E.g. the transition from a large number of chaotic and relatively small groups at the early stages of post-invasion resistance in Iraq since 2003 to fewer (but more consolidated ideologically and better organized) larger groups, mainly of the Islamicized nationalist bent, did not lead to less terrorism but resulted in more, better organized and more systemic terrorist activity.

[79] The period 1998–2006 is the only one for which the MIPT data on domestic terrorist incidents were available at the time of writing.

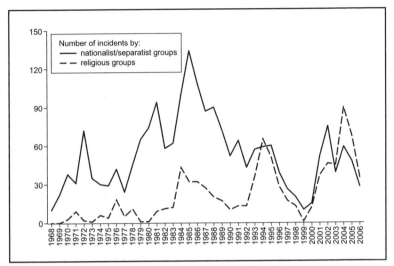

Figure 3.1. International terrorism incidents by nationalist/separatist and religious groups, 1968–2006

Source: MIPT Terrorism Knowledge Base, <http://www.tkb.org/>.

The comparative dynamics of key indicators—incidents, injuries and fatalities—for religious and nationalist/separatist terrorism at the domestic level over the period 1998–2006 are illustrated by figures 3.3–3.5. For all years of this period there were significantly more incidents by nationalist/separatist groups than by religious groups. The gap between the two only became narrower towards the end of the period. While in 1998 nationalist/separatist groups accounted for 3.7 times more domestic incidents than religious extremists, in 2006 it accounted for just 1.2 times more. Religious terrorism resulted in more injuries in domestic incidents in only three years (2003–2005) of the nine for which data are available. While religious terrorist groups caused more fatalities over the period than nationalist/separatist organizations, the latter accounted for more deaths in four years (1999–2002) out of the nine.

Thus, in terms of frequency of attacks, nationalist terrorism is understandably more widespread at the domestic level than religious terrorism. In terms of direct human costs—injuries and fatalities—the gap between religious and nationalist groups is narrower domestically

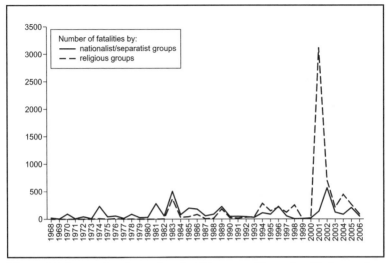

Figure 3.2. International terrorism fatalities caused by nationalist/separatist and religious groups, 1968–2006

Source: MIPT Terrorism Knowledge Base, <http://www.tkb.org/>.

than internationally, but religious terrorism in the early 21st century is generally more lethal, including at the domestic level.

As noted above, in terrorism research, sound conclusions cannot be reached on the basis of quantitative data alone and the rest of this chapter focuses on qualitative analysis. Nevertheless, it may be preliminary concluded from the analysis of quantitative data that, internationally, religious extremism has indeed become the most powerful motivational and ideological basis for groups engaged in terrorist activity. At the same time, the available data show not only that international terrorism lags behind domestic terrorism, in terms of both incidents and casualties, but also that, domestically, radical nationalism remains as powerful a mobilization tool for armed non-state actors as religious extremism.

Examples of violent extremism can be found in all large religions and in smaller confessions, religious currents and sects. Religious (and quasi-religious) terrorism may be associated with any religion and confession, and religious categories have been used to justify terrorist activity by groups of different religious or ethno-confessional orientations. These groups include the pseudo-Shinto Japanese sect

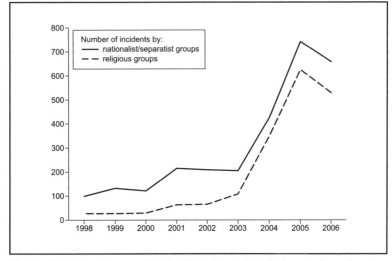

Figure 3.3. Domestic terrorism incidents by nationalist/separatist and religious groups, 1998–2006

Source: MIPT Terrorism Knowledge Base, <http://www.tkb.org>.

Aum Shinrikyo and radical Judaic, Hindu and Sikh extremists. However, in the late 20th and early 21st centuries the main terrorist threat to international security and to the security of many states—such as the USA and its Western allies, India, Russia, China and many Muslim countries—has been posed either by Islamist terrorism or by ethno-nationalist terrorism that has been Islamicized to varying degrees.[80]

---

[80] Almost half of the 42 organizations on the US State Department's list of foreign terrorist organizations (as of Oct. 2005) are Islamist groups. On the equivalent British list the proportion of Islamist groups is even higher, at 32 of the 43 international terrorist organizations proscribed as of July 2007. Russia's official list of terrorist organizations includes only groups that are either Islamist or Islamicized to some extent and all 4 groups on China's first list of terrorist organizations, published in Dec. 2003, are Islamicized 'Eastern Turkestan' separatist groups. US Department of State, Office of Counterterrorism, 'Foreign terrorist organizations (FTOs)', Fact sheet, Washington, DC, 11 Oct. 2005, <http://www.state.gov/s/ct/rls/fs/37191.htm>; British Home Office, 'Proscribed terrorist groups', <http://security.homeoffice.gov.uk/legislation/current-legislation/terrorism-act-2000/proscribed-terrorist-groups>; Borisov, T., '17 osobo opasnykh: publikuem spisok organizatsii, priznannykh Verkhovnym sudom Rossii terroristicheskimi' [17 most dangerous: groups listed as terrorist organizations by the Russian Supreme Court], *Rossiiskaya gazeta*, 28 July 2006, <http://www.rg.ru/2006/07/28/terror-organizacii.html>; and Xinhua, 'China identifies Eastern Turkistan terrorists', Beijing, 15 Dec. 2003, <http://news.xinhuanet.com/english/2003-12/15/content_1231167.htm>.

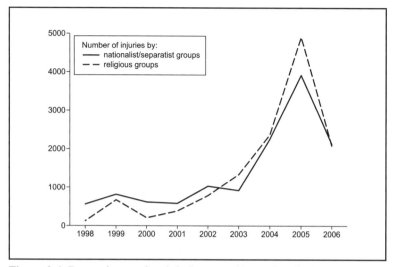

Figure 3.4. Domestic terrorism injuries caused by nationalist/separatist and religious groups, 1998–2006

Source: MIPT Terrorism Knowledge Base, <http://www.tkb.org/>.

When discussing the ideology of violent Islamism it is imperative to distinguish between religious and quasi-religious extremism. While the distinction may not always be strict and clear, it is most pertinent to the central issues of this Research Report. 'Purely' religious terrorism has been mainly practised by a limited number of marginal, closed religious groups and totalitarian sects. The religious extremism that provides the ideological basis for many broader movements usually goes far beyond religion and theology as such to encompass socio-political and socio-economic protest and issues of culture and identity.

Nowhere is this pattern more evident than in the case of violent Islamist extremism, including Islamist terrorism. Its quasi-religious character stems partly from the quasi-religious nature of Islam in its fundamentalist forms. Fundamentalist Islam provides a comprehensive concept of a social, political, ideological and religious order—a way of life and societal organization where religion, politics, state and society are inseparable. At the transnational level, the quasi-religious nature of radical Islamism is highlighted by the role it plays as an ideology of globalized violent anti-system protest. In playing that role,

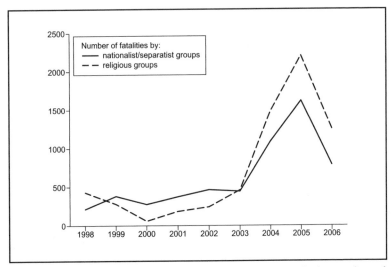

Figure 3.5. Domestic terrorism fatalities caused by nationalist/separatist and religious groups, 1998–2006

Source: MIPT Terrorism Knowledge Base, <http://www.tkb.org/>.

transnational violent Islamism has largely replaced the secular internationalist communist and leftist ideologies of the past. At a more localized level, the widespread combination of violent Islamism with various forms of nationalism and ethno-separatism also underscores, albeit in a different way, its quasi-religious character.

The links between religious radicalism and terrorism

The first major problem in studying the role played by religious radicalism in motivating, supporting, attempting to justify and guiding a certain group's terrorist activity is similar to the main theoretical issue raised in chapter 2 in relation to radical ethno-nationalism. The problem is that, while religious extremism may serve as a powerful driving force and may also be effectively instrumentalized to guide and justify terrorist activity, it does not necessarily or automatically lead to terrorism or, indeed, to violence.

In some national orientalist and Islamologist traditions a basic distinction is made between Islamic fundamentalism, primarily in its theological sense, and political Islamism. According to this interpretation, Islamic fundamentalism is practised by groups and individuals that may be very strict in scriptural terms but do not engage in political activism.[81] Political Islamism implies direct political action taken to advance fundamentalist goals. The prevailing view, however, seems to question this distinction as artificial. The term 'Islamic fundamentalism' is frequently used interchangeably with 'Islamism', with the latter being the preferable term to denote politically active and resurgent Islam.[82]

Whatever it is called, modern Islamism is a complex and multi-faceted phenomenon. Most commonly, it takes the form of broad reformist socio-political movements (often referred to as legalist Islamism). Despite their harsh criticism of and reservations about the existing order, movements such as most national branches of the Muslim Brotherhood or the Pakistan-based Jamaat-e-Islami are by and large ready to work within the system, principally in their own states, in order to change it.[83] More radical Islamism is represented by a set of extremist currents that are most commonly and directly associated with 'violent jihad' and are often—although not necessarily—engaged in violent activity. Thus, while in the late 20th and early 21st centuries Islamist terrorism has become the main form of transnational terrorism, Islamist movements and networks engage in a variety of activities dominated by different priorities. These may range from the socio-political to the missionary, with jihad (interpreted as a holy war against enemies of Islam) serving as the main priority for comparatively few groups.

The general distinction between mainstream (or legalist) and extremist Islamist actors is certainly useful, but the way in which it is commonly applied to the issues of violence and non-violence is a simplification. Mainstream Islamists are commonly associated with generally non-violent approaches, while all Islamist extremists at all levels are automatically linked to violence and especially to terrorism. The phrase 'extremists and terrorists' (as if these are always the two

---

[81] The distinction between theological fundamentalism and political Islamism has often been made by scholars of fundamentalist movements and is the prevailing approach in some national orientalist schools, notably in the Russian Islamologist tradition. See e.g. Malashenko, A., *Islamskoe vozrozhdenie v sovremennoi Rossii* [The Islamic renaissance in contemporary Russia] (Carnegie Moscow Center: Moscow, 1998).

[82] See e.g. the articles 'Fundamentalism' and 'Islamist' in Esposito, J. L. (ed.), *The Oxford Dictionary of Islam* (Oxford University Press: Oxford, 2003), pp. 88, 151.

[83] For more detail see section III below.

sides of the same coin), which has become common in all sorts of reporting and writing on the subject, is not fully accurate either. While all terrorists are extremists, extremists are not necessarily terrorists. Even some of the most professedly extremist, anti-system Islamist movements do not include armed jihad against their opponents as one of their main priorities and are not willing to use violence, especially against civilians. For instance, the Hizb ut-Tahrir movement in Central Asia, which originated from the broader, strongly extremist and transnationally active Hizb ut-Tahrir movement, has not just consciously opted to abstain from the use of terrorist means but has chosen non-violence in general.

The second problem in exploring the role of religious extremism in the instigation and ideological justification of terrorism is that groups using terrorist means in the name of religion do not necessarily represent some heretical, totalitarian 'deviant sects' or cults. Instead, they are often guided by a radical interpretation of their religion's basic concepts, such as the radical militant interpretation of the essential Islamic notion of jihad. The ideologues of such groups tend to argue that, on the contrary, it is the moderate majority of the clergy and ordinary believers that has deviated from the basic tenets of the faith and call for a return to what they see as its untainted beliefs, values and practices. The long road of 'return', or revival, would imply stricter observance of the 'original' religious rules (which is precisely what most religious fundamentalists opt for). Extremists, however, tend to promote and follow a much shorter road of 'purification' though violence and self-sacrifice (i.e. suicide) in the course of the 'holy war'.

Third, while some generalizations are possible in the analysis of the relationship between religious extremism and terrorism, they should be applied with extreme care. This care is required in view of the specific features of terrorism supported and inspired by different versions of religious extremism. It is also dictated by the wide variety of groups, movements and currents that may be associated with the 'radical' wing of the same confession. While these may be covered by the same term, such as 'Islamism', only a few resort to violence, let alone terrorism.

II. Similarities and differences among violent religious and quasi-religious groups

Certain general features are shared by most terrorist groups that are guided by an ideology with a strong religious imperative.

First, for such organizations and movements the use of terrorist means (and especially significant, large-scale or mass-casualty attacks) usually necessitates formal blessing by some spiritual authority or guide. These spiritual leaders may hold a senior or leading position in the organization or may be independent of it.[84] For Islamist terrorists, the formal blessing usually takes the form of a special religious and legal pronouncement (fatwa[85]), which legitimizes the use of terrorist means and may either precede or follow the act of terrorism.[86]

In fact, one of the main formal criteria for identifying an armed group as the one whose ideological basis is predominantly religious is precisely the presence of clerical figures in a group's leadership. This is especially so if this presence is combined with the consistent use of religious rituals or sacred texts for the inspiration and justification of violence, including terrorism, and for activities such as attracting and recruiting new members. This extends to movements with multiple leaders and networks with an even more dispersed, diversified or even 'virtual' leadership—a pattern that characterizes the post-al-Qaeda transnational violent Islamist movement. Such a movement may be associated with different types of religious leaders, scholars and clerics; for example, the old generation of al-Qaeda scholars and the new Internet generation of 'jihadi scholars'.

The more politicized a group and the broader the range of its functions are, the more likely it is that at least some of its spiritual guides are based outside its formal organizational framework. For example, when taking important decisions, including those concerning terrorist

---

[84] Examples of spiritual guides with a leading position within their group include the late Ahmed Yassin, founder of Hamas, Muhammad Hussein Fadlullah and Hassan Nasrallah of Hezbollah, the Sikh leader Jarnail Singh Bhindranvale and the pseudo-Shinto 'messiah' Shoko Asahara of Aum Shinrikyo.

[85] A fatwa is an opinion or ruling on Islamic law (sharia), traditionally made by highly esteemed Islamic scholars to settle difficult or unclear cases.

[86] See e.g. Lakhdar, L., 'The role of fatwas in incitement to terrorism', Special Dispatch Series no. 333, Middle East Media Research Institute (MEMRI), 18 Jan. 2002, <http://memri.org/bin/articles.cgi?Page=archives&Area=sd&ID=SP33302>.

activity, the Hamas leadership may specifically consult Islamist theologists and spiritual authorities outside the Palestinian territories.[87]

In some cases, the spiritual leaders of a group do not have solid theological credentials or clerical education. This usually points to the group's quasi-religious, rather than purely religious, character—that is, to its goals and agenda being highly politicized. The best-know example is, of course, bin Laden, who lacks any proper theological credentials, education or reputation but effectively poses as a spiritual leader and an oracle for the Muslim world. By issuing fatwas, he has used an Islamic religious and legal instrument to convey what are essentially political manifestos.

Second, groups that are indeed guided by a strong religious imperative, as opposed to organizations that are merely formed on an ethnoconfessional or sectarian basis, tend to explicitly justify armed violence, including terrorism, by making direct references to sacred texts. These texts are not necessarily apocryphal or heterodox but may include the holy books or traditional writings that are fundamental to a certain religion or confession, such as the Quran or the Hadith for Islam.[88] Not surprisingly, different extracts from the same texts may be employed by more moderate forces to justify exactly the opposite point.

Third, both religious and quasi-religious terrorist groups do not limit themselves to the use of sacred texts. They actively employ and adjust religious and quasi-religious rituals and cults, such as self-sacrifice and the cult of martyrdom, for their purposes. In this way, those who carry out a terrorist act see themselves and are perceived by their group and its supporters as martyrs for faith. In contrast to many secular militant organizations, for terrorist groups whose ideology is strongly influenced by religious extremism, the upgrading of a terrorist attack to an act of faith (especially when carried out as an act of self-sacrifice) effectively removes some of the basic constraints on incurring mass casualties. It thus facilitates the perpetration of deadlier, large-scale attacks.

---

[87] E.g. Hamas frequently uses the fatwas of Qatar-based Yusuf al-Qaradawi as source of religious authority and posts them on its website. See e.g. Middle East Media Research Institute (MEMRI), 'Sheikh al-Qaradhawi on Hamas Jerusalem Day online', Special Dispatch Series no. 1051, MEMRI, 18 Dec. 2005, <http://memri.org/bin/articles.cgi?Page=archives& Area=sd&ID=SP105105>.

[88] The Hadith are narrations about the life, actions and sayings of the Prophet Muhammad.

No less important is the active use of religious symbols and images and the interpretation of political reality through these symbols. While religious symbols and images employed by violent groups may be as basic and archetypal as are those used by nationalists in their myths, they tend to be even more abstract and generalized. Even if they are personalized—that is, associated with some specific political or religious figures—these images become symbolic of the 'heroes' or 'enemies' of faith and by definition become universalized by acquiring sacred meaning.

In particular, religious extremists identify, interpret and see 'the enemy' in much broader, almost universal terms than do secular groups or ethno-confessional groups that do not emphasize religion. The enemies may be personalized to some extent by certain key figures, but they are used as examples of the more generalized notion. For instance, the standard calls by radical Islamist scholars associated with the post-al-Qaeda movement are to 'fight all the infidels, whether apostates or Crusaders, nationals or foreigners, Arabs or non-Arabs'. But they may specify the enemies: 'their names be Abd al-Aziz Bouteflika, Abdallah bin Abd al-Aziz, Abdallah bin Hussein, Mu'ammar Qadhafi, or George Bush, Tony Blair, Sarkozy, or Olmert'.[89] Ultimately, however, the enemy cannot be reduced to a handful of individuals (as used to be the case for socio-revolutionary terrorists of the late 19th century). Nor is it limited to a certain social class or ethnic group (as is often the case for modern leftist or ethno-nationalist radicals). Rather, the ultimate enemy is likely to represent some generalized and impersonalized evil, a ubiquitous Satan. In other words, for terrorists guided by a strong religious imperative, the main protagonist can only be defined in very broad and rather blurred religious (political, ideological, politico-geographic) categories. The enemy may, for example, range from the West to the entire world of unbelief, ignorance and materialism (*jahiliyyah* in Islam) or 'all injustice on earth'.[90]

---

[89] Quoted from a July 2007 statement by radical Islamist Internet generation scholar Abu Yahya al-Libi as translated by the Global Research in International Affairs (GLORIA) Center, Project for the Research of Islamist movements (PRISM) in Paz, R., 'Catch as much as you can: Hasan al-Qaed (Abu Yahya al-Libi) on Jihadi terrorism against Muslims in Muslim countries', PRISM Occasional Papers, vol. 5, no. 2 (Aug. 2007), <http://www. e-prism.org/projectsandproducts.html>, p. 4.

[90] *Jahiliyyah* is a traditional Islamic notion referring to the state of lawlessness and ignorance in the pre-Islamic period; it literally means 'ignorance' in Arabic and is used to denote ignorance of divine guidance. It is also employed by radical Islamists to denote the current

Against this background, it is unsurprising that quasi-religious Islamist ideology has emerged as the replacement for the secular radical socio-revolutionary ideas of the past as the main justification of the type of modern terrorism that goes beyond localized contexts. Transnational Islamist ideology is most effective at playing this role for superterrorism with a global reach and agenda.

Fourth, while the demonstrative effect of a terrorist attack and terrorism in general on a particular state, a group of states, or domestic or international public is important, the main audience for terrorists guided by strong religious imperative tends to be a witness of a much higher order. For Islamist terrorists in particular, 'Allah sufficeth as a Witness'.[91] A terrorist act, especially one that involves self-sacrifice is also important for the religious or quasi-religious terrorist himself. Such an individual or collective ritual act is directed to the terrorist himself and his religious associates to no less an extent than to the enemy that is to be impressed and terrorized.

Finally, most religious, and especially quasi-religious, armed extremist groups (regardless of their confession) do not as a rule draw a clear distinction between religion and politics. This trend is most developed in Islamist organizations, both those that do not use violence and those that engage in violent activity. This is in large part due to the holistic, all-embracing nature of Islam, where legal and normative aspects of life are developed in far greater detail than in other religions. In that sense, Islam, especially in its fundamentalist forms, is more of a comprehensive concept of social order and organization, at both the national and supranational levels, than other religions and confessions. Islamist opposition groups, in particular—both legalist movements and more radical violent organizations—have long used religious discourse to embrace a broad range of essentially political, social and economic demands.

A combination of all these characteristics helps to distinguish between, on the one hand, groups for whom religion is nothing more than an essential part of their ethno-confessional background and, on

state of unbelief, ignorance and materialism in the world that is not governed by norms of fundamentalist Islam. See e.g. Qutb, S., *Milestones* (Unity Publishing Co.: Cedar Rapids, Iowa, 1980), pp. 11–12, 19–22, 56 etc. See also section IV below.

[91] Quran, sura 48, verse 28, transl. Muhammad Pickthall. This can also be translated as 'Allah is enough for a Witness' (transl. M. H. Shakir). Translations of the Quran by these 2 translators and by A. Yusufali are taken from *The Noble Qur'an*, University of South California, Muslim Student Association, Compendium of Muslim Texts, <http://www.usc.edu/dept/MSA/quran/>.

the other hand, genuinely religious extremist groups. In this context, a dictum by François Burgat, a leading French expert on Islam, stating that 'The Quran can "explain" Osama bin Laden no more than the Bible can "explain" the IRA' is, with due respect, not very accurate.[92] While all IRA members are Catholic, the group (in contrast, for instance, to Hamas or al-Qaeda) did not systematically employ religious sermons or quote from sacred texts to justify armed violence. Nor did the IRA require clerical authorities to sanctify violence and, specifically, the use of terrorist means.

Along with the above common features shared by most terrorist groups with a strong religious imperative, there are multiple and major differences among them in terms of structure, scope and types of activity. The most basic distinction can be made between totalitarian religious sects (such as the pseudo-Shinto Aum Shinrikyo or the US-based radical Christian movements) and religious and quasi-religious groups of all other types.

For instance, while totalitarian messianic sects and cults have very strict hierarchies, religious and quasi-religious groups of other types are very diverse in their organizational forms. The latter groups may range from broad religious, social and political grass roots movements to small radicalized cells that have split off from larger, usually more moderate, movements and communities. Other than the strictly hierarchical totalitarian sects, most groups guided by a religious imperative tend to be more loosely structured than, for instance, ethno-nationalist organizations. Violent Islamist groups and movements, especially those active at the transnational level, appear to have the most flexible, fragmented, networked yet surprisingly well coordinated structures. Their semi-autonomous multiple cells constantly adapt themselves to the environment, resurface and interact in various combinations and reorganize themselves.[93]

It should also be kept in mind that, in contrast to al-Qaeda and the post-al-Qaeda transnational Islamist movement, most groups that operate locally and are Islamist or have become Islamicized effectively combine religious extremism with radical nationalism. This is the case in Chechnya, Iraq, Kashmir or Mindanao. This means that both the ideologies and structures of such groups are affected by the specific local contexts, multiple—ethnic, tribal, regional and

---

[92] Burgat, F., *Face to Face with Political Islam* (I. B.Taurus: London, 1997), p. xv.

[93] On organizational patterns of terrorist groups see chapters 4 and 5 in this volume.

national—cultures and identities and other characteristics. This makes the range of groups that employ religion as an ideological basis for terrorist activity even less homogeneous.

The ideological and structural diversity of these violent groups may also be demonstrated by their different degrees of involvement in politics or social and humanitarian work. In addition, there are different approaches towards apostates who used to be active members of a militant organization but have left or split from it. While in totalitarian religious sects such a betrayal is often punished by death, for Islamists, for instance, it does not necessarily pose a major problem from an organizational point of view. Despite multiple splits and feuds, the high degree of structural flexibility and fragmentation in Islamist networks, such as the Jemaah Islamiah (JI) network in South East Asia, enables them to form new alliances with the splinter groups and gives the former apostates a chance to rejoin the movement. This is in line with the principle that the best way to repent for someone who has betrayed 'jihad' is to wage 'jihad'.[94] These ideological and structural principles help maintain overall organizational stability and sustainability despite constant splits, regroupings and transformations.

III. Terrorism and religion: manipulation, reaction and the quasi-religious framework

There are different analytical approaches to the role that the ideology of religious extremism plays in the justification, sanctification, motivation and ideological support of terrorism. Most of them can be categorized by their emphasis on either pragmatic manipulation or broader reaction to social, political, identity and other factors. While the first approach is focused on the terrorists' manipulation of religion for political purposes, the second approach views religious extremism itself as a form of genuine socio-political protest.

---

[94] On how this principle is applied in the case of JI see International Crisis Group (ICG), *Recycling Militants in Indonesia: Darul Islam and the Australian Embassy Bombing*, ICG Asia Report no. 92 (ICG: Brussels, 22 Feb. 2005), <http://www.crisisgroup.org/home/index. cfm?id=3280>, p. 7.

Manipulation

Those analysts who emphasize various manipulative and instrument-alist interpretations try to address the problem in a more applied sense. These are mostly analysts or political commentators who specialize in security issues, including terrorism, but lack expertise in Islam and Islamism per se. They argue that Islamist and other religious extrem-ism is simply manipulated for political purposes by terrorist groups and especially by their leaders and ideologues.[95] Indeed, it should be recognized that religious extremism can be effectively instrumental-ized to some extent in terrorism-related contexts for several reasons.

First, religious extremism provides both a convenient means to communicate a political message and a ready-made information system. This system of well-established channels of communication is formed by a network of religious study groups, associations, institu-tions, publications, websites, Internet blogs and forums and so on. It enables a terrorist group to convey its message in a religious form, including by formal religious legal rulings. This communicative advantage of framing a message in religious form and discourse allowed bin Laden and some other leaders of the transnational violent Islamist movement, who were not recognized clerical figures, to issue essentially politico-military manifestos in the form of fatwas.[96] This approach can be summed up as follows. The terrorists' message may not necessarily be explicitly religious, but they skilfully use a reli-gious form to deliver this message both to 'the enemy' and to as broad an audience as possible and to give it additional power of persuasion.

Second, socio-political, ethno-nationalist and other resentment may often be channelled into religious discontent. This resentment is then articulated in religious categories and discourse. An additional advan-tage of channelling socio-political and especially ethno-nationalist resentment into religious form is that it may effectively help to trans-nationalize a group's agenda and broaden its constituency. In the vac-uum that results from a lack of equally powerful secular ideologies in

---

[95] In the early 21st century, one of the most prominent proponents of this view has been Bruce Hoffman. See e.g. 'Religion and terrorism: interview with Dr. Bruce Hoffman', Religioscope, 22 Feb. 2002, <http://www.religioscope.com/info/articles/003_Hoffman_terror ism.htm>. This is in contrast to his earlier views of 'religious terrorism' interpreted as vio-lence that is first and foremost a 'sacramental act or divine duty executed in direct response to some theological demand or imperative'. Hoffman, '"Holy terror"' (note 77), p. 272.

[96] See also section II above.

the early 21st century, resort to religious extremism allows terrorists to extend their constituency far beyond, for instance, members of a certain ethnic group. Instead, they can appeal to an audience of many millions within a broader religious community. There their message can receive much broader support, even if their tactics are rejected by the majority of that community.

It cannot be denied that a certain degree of manipulation of the religious factor by modern terrorists and especially by their leaders and ideologues along the lines described above does take place. Overall, however, this approach tends to significantly simplify the link of terrorism with religion and, more specifically, religious extremism. It ignores or downplays a set of objective socio-political and cultural changes in the Muslim world that are going on under the multiple pressures of modernization, globalization and Westernization. These pressures reinforce (and are themselves aggravated by) the perception among many Muslims of the essentially anti-Islamic nature of the policies of the USA, other Western states and 'impure', corrupt, Westernized and elitist regimes in many Muslim countries. This vision is also reinforced by the perceived long history of repression and suppression of Muslims by colonial powers, secular nationalist regimes, and so on. Finally, the serious problem with the approach that emphasizes a manipulative connection between terrorism and religion is that it almost by definition denies terrorist groups and their leaders genuine religiousness and religious conviction.

Reaction

In contrast, such prominent scholars as François Burgat and John Esposito, while they differ in their explanations of Islamism, agree that it has more fundamental roots and a broader role to play. The role of religious radicalism is seen by these and other scholars of Islam as a reaction of part of the disillusioned elites and societies in the Muslim world to some painful social and socio-political realities associated with traumatic modernization, secularization and Westernization.[97] More specifically, it is also a reaction to the dominant pat-

---

[97] This includes the prevalence of 'corrupt authoritarian governments and a wealthy elite ... concerned only with its own economic prosperity, rather than national development, a world awash in Western culture and values'. Esposito, J. L., *Unholy War: Terror in the Name of Islam* (Oxford University Press: Oxford, 2002), p. 27.

terns of political violence in their own societies and to certain policies of international actors, who are perceived as meddling, aggressive and anti-Muslim. Some authors go even further, arguing that Islamist activism in particular is not merely a reactive force but incorporates elements of genuine socio-political protest that are more in line than in conflict with the drive toward modernization.[98]

It may be added by the present author that Islamism as a reaction to these broad social, political and cultural processes is to some extent inevitable and in this sense is close to being a reflex or symptom. Even if some of the ideologues of violent Islamism refuse to see their own actions as defensive and reactive, all Islamist terrorist groups tend to become active in an environment that they perceive as a crisis, or even as catastrophic. They see these crisis conditions as threatening the identity or physical survival of their social or ethno-confessional group or of a much broader community, such as the entire Muslim *umma*. They also effectively build on real grievances based on past and present injustices (such as the US-led interventions in Afghanistan and Iraq) committed by Westernized 'modernists', 'aliens' or 'non-believers' against a community in whose name terrorists claim to speak and act.

The reactive character of violent Islamist extremism, especially when it reaches the point of resort to terrorism, is most evident wherever there is something against which to react. The rise, radicalization and militarization of Islamist groups and movements is most common in areas of the closest contact with different political, governance, socio-economic and value systems. Those points of contact range from Muslim diasporas in Western countries to areas of visible Western presence in the Muslim world. Some of the most problematic cases are those of the US-led interventions in Afghanistan and Iraq, the US military presence in the Arab states of the Gulf and the Israeli occupation of Palestinian territories. The points of contact may also include those Muslim states that have themselves been most affected by rapid, uneven, particularly painful and traumatic modernization and secularization (e.g. Egypt). These trends further widen the gaps between the bulk of the population in those countries and the relatively secularized elites and between modern and traditional ways of life.

---

[98] According to Burgat, Islamism may thus even pose as a progressive force in conservative Islamist clothing. See e.g. Burgat (note 92), pp. xiii, xvi, 165–166, 179.

Needless to say, analytical approaches that follow this general direction and are rooted in Islamic and orientalist studies, socio-ideological analysis or political sociology are more accurate and adequate in their analysis of Islamism. Unfortunately, they too fail to provide a full explanation of the phenomenon of Islamist terrorism. It even seems that the more the world's leading academic experts on Islam are interested in broader social, political, cultural and identity aspects of Islamism, the less specific interest they show in Islamist violence, including terrorism. They seem to be more likely to dismiss different forms of such violence, especially terrorism, as mere excesses of an extremist fringe and to disregard the entire phenomenon as simply marginal.[99] Also, their vast knowledge of Islam and Islamism in broad comparative national and cultural contexts is rarely matched by an equal degree of familiarity with terrorism as a specific tactic of political violence. For example, terrorism is often confused with other modes of operation or with violence in general.[100] In other words, those academics who are best at explaining Islamism have problems with, or show little interest in, explaining Islamist terrorism.

However, the phenomenon of Islamist terrorism, especially in its transnational forms that are not confined to any specific local context, needs to be explained. It cannot simply be dismissed or ignored, if only because this 'extremist fringe' has managed to attract disproportionately high political attention to its programme and goals. In media and public discourse it has effectively managed to outmatch the multiple varieties of mainstream political Islam. As a result, in terms of political impact, cells of the transnational Islamist movement employing terrorist means can be more accurately described as an 'overwhelming minority' than as a 'marginalized minority'. Also, while it is not just the Islamist fringe that uses violence in Muslim states and regions, the kind of violence that is the focus of this Research Report—asymmetrical, mostly indiscriminate terrorism against civilians—is the tactic dominated by fringe actors.

---

[99] Burgat (note 92), pp. xvi, 167, 178.

[100] Most Islamologists tend not to define terrorism and many often use 'terrorism' as a synonym for 'violence'. This leads them to all sorts of confusion, such as references made to 'terrorism in early Islam'. E.g. Esposito (note 97), pp. 29, 36, 41. In contrast, experts specializing in terrorism attribute the emergence of 'terrorism' as a specific tactic of politically motivated violence to the second half of the 19th century at the earliest and usually distinguish it from the broader term 'terror', which has a longer history.

The quasi-religious framework

If the world's most sophisticated scholars of Islam keep under-estimating or de-emphasizing Islamist violence in the form of terror-ism, there should be no surprise that the vacuum is filled by studies of a speculative nature and unsatisfactory quality. If serious Islam-ologists, political sociologists and cultural anthropologists do not come up with a thorough explanation for Islamist terrorism, the field will continue to be dominated by security experts.

One way to overcome this problem is to supplement the broader socio-ideological and politico-sociological approach to Islamism and Islamist violence with an emphasis on the quasi-religious nature of modern violent Islamism. There are two possible extremes—both of which should be avoided—in interpreting the quasi-religious nature of Islamism in general and violent Islamism in particular.

One extreme is to stress the 'religious' part of the term 'quasi-religious. This approach, which largely reduces Islamist terrorism to purely religious terrorism, is still popular among Western analysts. This view de-emphasizes the fact that the violent Islamists' demands are never just theological and are equally, if not more, political. Nor does it sufficiently take into account the radical Islamist interpretation of religion itself, according to which 'Religion means the system and way of life that brings under its fold human life with all its details'.[101] This interpretation goes beyond the standard contemporary Western understanding of religion and its role in society.

In this context, it is important to note that the end state of Islamism, both violent and non-violent—an Islamic Caliphate that is ultimately to spread all around the world[102]—is by no means an analogue of a theocratic state in its Western interpretation. The global Islamic Caliphate is not an Islamist version of the Roman Catholic Vatican. Rather than the rule of a clerical hierarchy, the Caliphate is supposed to embody the 'direct rule of God'. Finally, even the most radical Islamist ideologues advocating violence in the name of 'jihad' concur that, while an Islamic order can be imposed by force, this is not the same as imposing Islam as a religion by force. It is accepted that Islam

[101] Qutb, S., 'War, peace, and Islamic Jihad', eds M. Moaddel and K. Talattof, *Contemporary Debates in Islam: An Anthology of Modernist and Fundamentalist Thought* (Macmillan: Basingstoke, 2000), p. 231.

[102] Caliphate is the Islamic form of government based on the 'direct rule of God' and uniting all Muslims. The Ottoman Empire is considered to be the 'last Caliphate'.

'forbids imposition of belief by force, as is clear from the [Quranic] verse, "There is no compulsion in religion"'.[103] In the Islamist tradition, the mere fact of someone having a different belief has generally not been seen as a sufficient cause for violent jihad. As stated by Sayyid Abul Ala Maududi, founder of Jamaat-e-Islami, 'the objective [of jihad] is not to coerce the opponent to relinquish his principles but to abolish the government which sustains these principles'.[104]

An opposite extreme is to stress the 'quasi' element in violent quasi-religious Islamic extremism. This view stresses the essentially political anti-Western, anti-imperialist and anti-neocolonialist goals of violent Islamism while de-emphasizing any specific religious motivation and mobilization power. This bias is not limited to those who emphasize the manipulative aspect of quasi-religious extremism and deny its religious aspects any more significant role than that of a form or communication channel. It may also be shared by some of those scholars who focus on the reactive nature of the link between religious extremism and violence but underestimate the power of ideology and religious doctrine and belief to influence 'the forms of action favoured by Islamist movements'. These forms of action are seen as 'directly inspired by dominant political actors at both local and international levels'. At the transnational level, violent Islamism is interpreted as little more that a protest against 'the current impasse in relations between a large part of the Muslim world and the West—especially the US'.[105]

In fact, as shown in the rest of this chapter, the dialectic nature of modern violent Islamism implies a combination of genuine religiousness with the use of a religious framework for communication and to justify and legitimize political–militant action and goals. The central importance of a high degree of religious conviction and of the notion of faith (*imaan*) as an ultimate test, goal and standard of armed struggle should not be underestimated.

---

[103] Qutb (note 101), p. 227. The quote from the Quran is sura 2, verse 256.

[104] Maududi, S. A. A., 'Jihad in Islam', Lecture given in Lahore, 13 Apr. 1939, reproduced in Laqueur, W. (ed.), *Voices of Terror: Manifestos, Writing and Manuals of Al-Qaeda, Hamas, and Other Terrorists from around the World and throughout the Ages* (Reed Press: New York, 2004), p. 400.

[105] Burgat (note 92), pp. xiii, xv. This view is summarized by Burgat's call to rely on 'political sociology', rather than 'holy books', in trying to understand Islamism, including violent Islamism. Burgat (note 92), p. 8. In contrast, the argument made in this study can be summarized as 'both are essential'.

IV. The rise of modern violent Islamism

An analysis of contemporary violent Islamism requires at least a brief historical overview. There is no need to retell the history and pre-history of modern Islamism in great detail, as it has been done well elsewhere by orientalist and Islamic studies scholars.[106] However, several landmark developments should be mentioned.

The Islamic fundamentalist movement (Salafism) has its roots in the painful reaction of the Muslim world, and especially of its Sunni part, to the fall of the 'last Caliphate'—the Ottoman Empire—following the end of World War I.[107] Over the following decades, the theory and practice of Islamism—political activity to advance the fundamentalist agenda, with the re-establishment of the Islamic Caliphate as the rhetorically ultimate goal—started to develop. It is important to stress that the early Islamist movements were non-violent. They gave rise to a moderate current in Islamism (now often referred to as legalist), while the more radical current took longer to form.

The reactive nature of Islamic fundamentalism and Islamism in response to painful, ineffective and elitist modernization (which is primarily blamed by Islamists on Westernization) should be kept in mind. At the same time, the power of secularized modernization in the post-World War II period—and of nationalist, left-wing and other ideologies associated with it—in socio-political, economic and cultural spheres should not be underestimated. Throughout the 20th century most anti-colonial movements in the Arab world were guided by secular nationalist ideologies. Examples include Nasserism in Egypt,

---

[106] In addition to works cited above, suggested background texts include: Ayoob, M. (ed.), *The Politics of Islamic Reassertion* (St. Martin's Press: New York, 1981); Ayubi, N. N., *Political Islam: Religion and Politics in the Arab World* (Routledge: London, 1991); Roy, O., *The Failure of Political Islam* (Harvard University Press: Cambridge, Mass.,1994); Esposito, J. L. (ed.), *Political Islam: Revolution, Radicalism or Reform* (Lynne Rienner: Boulder, Colo., 1997); Tibi, B., *The Challenge of Fundamentalism: Political Islam and the New World Disorder* (University of California Press: Berkeley, Calif., 1998); Rubin, B. (ed.), *Revolutionaries and Reformers: Contemporary Islamist Movements in the Middle East* (State University of New York Press: Albany, N.Y., 2003); Wiktorowicz, Q. (ed.), *Islamic Activism: A Social Movement Theory Approach* (Indiana University Press: Bloomington, Ind., 2004); and Keppel, G., *The Roots of Radical Islam* (Saqi: London, 2005).

[107] The Ottoman Caliphate was formally abolished by Turkish President Mustafa Kemal Atatürk in 1924.

Baathism in Iraq and Syria, the Neo-Destour movement (or Bourguib-ism) in Tunisia, as well as the FLN in Algeria and the PLO.[108]

The moderate, legalist current in modern Islamism can be traced back to the emergence of two organizational networks. One is the Jamaat-e-Islami movement, founded in 1941 in British-ruled India by Maududi and now based in Pakistan. The other is the Muslim Brother-hood movement that was established by Hassan al-Banna in Egypt in the late 1920s and early 1930s and was actively opposed to secular Nasserism in the post-World War II period. Subsequently, many other Islamist groups were formed within these broad movements or in association with them. Jamaat-e-Islami and the Muslim Brotherhood called for the gradual transition to Islamic rule and the creation of Islamic states through peaceful means as an alternative to secular, Western-style socio-political development and modernization. In theological terms, these moderate Islamists were in many ways close to Saudi Wahhabism, which forms the basis for the Islamic state in Saudi Arabia and its structures such as the Council of Senior Ulema.[109] Jamaat-e-Islami and most branches of the Muslim Brother-hood, which are active in many countries, have effectively combined religious reformism with political mobilization. In modern Egypt, for instance, legalist Islamists represent the only mass-based political movement.

---

[108] Nasserism is a secular pan-Arab socialist nationalist ideology (Arab socialism) associated with the name of Egyptian President Gamal Abdel Nasser (president, 1954–70).

Baathism is another version of a pan-Arab nationalist, Arab socialist ideology. The Baath Party was founded in Syria in 1947 and a branch was established in Iraq in 1954. It came to power in both countries in 1963. Baathists remain in power in Syria but were deposed in Iraq in 2003 by the US-led invasion.

The Tunisian Neo-Destour (New Constitution) Party succeeded the nationalist Destour Party in 1934 as a secular, modernist national liberation movement against French colonial rule. It was founded and led by Habib Bourguiba, who became the first president of independent Tunisia in 1957. For a brief period, from the mid-1960s until the early 1970s, the movement experimented with socialism.

On the FLN see chapter 2 in this volume, section II.

The PLO—a multi-party Palestinian political confederation of a nationalist and mostly secular character—was founded in 1964 as a national liberation resistance movement. See also chapter 2, section II.

[109] Wahhabism is the movement of followers of Muhammad ibn Abd al-Wahhab (1703–92), who called for the 'purification' of Islam and the revival of its 'original' version that should strictly replicate the way of life of the Prophet Muhammad and the first generations of Muslims. Wahhabism has been the official form of Islam in Saudi Arabia since the kingdom's creation in 1932. The Council of Senior Ulema (or Higher Council of Ulema) is a body for regular consultations between the monarch and Saudi religious leaders (*ulema*) created in 1971.

Sayyid Qutb, an Egyptian school inspector who was one of the founders of modern radical Islamism,[110] not only developed but also revised some of al-Banna's ideas. Qutb became one of the main ideologues of contemporary violent Islamism (commonly, but not entirely correctly, referred to as 'jihadi Islamism').[111] While other radical thinkers contributed to developing this trend, he stands out as the author of the most comprehensive, intellectually coherent and professedly extremist interprêtation of violent jihad. His powerful message inspired many radicals in his own time, but has truly resonated decades later with the emergence of al-Qaeda and the post-al-Qaeda movement of the late 20th and early 21st centuries.

Qutb characterized modern society as being 'steeped in *Jahiliyyah*' that, like a 'vast ocean', 'has encompassed the entire world'.[112] He considered the new *jahiliyyah* of the modern age to be much worse than 'the simple and primitive form of the ancient *Jahiliyyah*' that preceded the coming of the Prophet Muhammad.[113] Qutb regarded any society as being part of *jahiliyyah* as long as it 'does not dedicate itself to submission to God alone, in its beliefs and ideas, in its observances of worship, and in its legal regulations'. Not surprisingly, no existing societies met this definition, which for Qutb meant that they were all *jahili*.[114]

Qutb's earlier interest in socialism was reflected in a clear social connotation to his interpretation of *jahiliyyah*. Among its essential characteristics he listed exploitation, social injustice, oppression of the poor majority by the rich minority and tyranny. According to Qutb, *jahiliyyah* rots morale and spreads like a disease so that the people may not even suspect that they are 'infected'.

Qutb regarded Western society in particular as materially prosperous but morally rotten, 'unable to present any healthy values for the guidance of mankind' and possessing nothing that 'will satisfy its own

---

[110] The historical roots of radical Islamism date back to the 13th and early 14th centuries, with Ahmad ibn Taymiyyah (1263–1328) considered to be one of its spiritual fathers. See e.g. Esposito (note 97), pp. 45–46.

[111] On Qutb's return from a trip to the USA, where he studied in the late 1940s, he joined the Muslim Brotherhood, was arrested for his opposition to the Nasser regime and was executed in 1966 on charges of attempting to overthrow the secular Egyptian Government.

[112] Qutb (note 90), pp. 10, 12.

[113] Qutb (note 90), p. 11.

[114] Qutb (note 90), p. 80.

conscience and justify its existence'.[115] However, what distinguished Qutb is that he not only fully recognized the power of modernization but was also quite rational in supposing that *jahiliyyah* might prevail over Islam. Qutb saw people as being slaves to material benefits and animal instincts, which form the main essence of the modern unspiritual world, and as having no intention or motivation to counter or restrain these instincts.

According to Qutb, this problem can only be solved by creating a society of a new type that under the leadership of Islamists will be able to establish and sustain the moral framework required to successfully confront *jahiliyyah*. The powerful moral imperative and social accent in Qutb's conceptual thinking are further reinforced by ideas that may even appear to be reminiscent of anarchism, especially inasmuch as he rejects state power. In this interpretation, Islam is viewed as 'a universal declaration of man's freedom from the servitude to other men'.[116] It strives 'to annihilate all . . . systems and governments that establish the hegemony of human beings over their fellow beings and relegate them to their servitude', and for an 'all-embracing and total revolution against the sovereignty of man in all its types, shapes, systems, and states'.[117]

Remarkably, well before the debates on globalization emerged, Qutb's Islamism in many ways had already presented an alternative version of globalization. Essentially, it advanced its own vision of supranational globalism. This type of globalism is based on and ruled by Islam but provides for cultural and ethno-confessional pluralism (on condition that their adherents recognize the primacy of the 'One God'). All this has made Qutb's Islamism not just an extremist theory, but also a powerful and a surprisingly modern ideology that gives a radical, fundamentalist response to the challenges of the modern world.

While rejecting any possibility of a compromise between Islam and *jahiliyyah*, or between God and Satan, Qutb was fully aware of the difficulties that the fight against *jahiliyyah* entails. He was particularly

---

[115] Qutb (note 90), p. 7. It is interesting to note that Qutb also denounced the Marxist ideology of the Communist states. While recognizing that many people were attracted to Marxism as 'a way of life based on a creed', he argued that 'This ideology prospers only in a degenerate society or in a society which has become cowed as a result of some form of prolonged dictatorship'. Qutb (note 90), p. 7.

[116] Qutb (note 101), p. 227; see also p. 231.

[117] Qutb (note 101), pp. 228, 231.

sceptical about the chances for getting any significant support from the 'passive public masses' in this undertaking. Thus, he put forward an idea of a vanguard elite protest movement. It should lead the masses to realization of the 'Supreme Truth' through the 'revolution from above', carried out through a variety of means, including armed jihad.[118] In many ways, this vision forestalled the emergence of the loosely organized al-Qaeda movement over the 1990s. It can even more accurately describe the more fragmented and looser network of semi-autonomous or fully autonomous cells of the post-al-Qaeda movement of the early 21st century.

In Qutb's opinion, this vanguard takes on the mission of reviving Islam, ending the power of man over other men and establishing the rule of God. In order to achieve its mission, it should separate itself from the 'impure' environment, 'become independent and distinct from the active and organized *jahili* society whose aim is to block Islam'.[119] It is in line with this principle that the multiple autonomous cells of the contemporary transnational violent Islamist movement are formed. It is striking that even if the members of the modern Islamist cells inspired by al-Qaeda are not necessarily familiar with Qutb's writings, the cell-formation process tends to follow the precepts that he laid out.

The rise of violent Islamism, in both theory and practice, was also prompted and facilitated by a set of political and politico-military developments of the late 1970s and early 1980s. The anti-Shah revolution in Iran in 1979–80 for the first time proved that a mass-based Islamist movement could come to power through violent means. Following the anti-Soviet 'jihad' in Afghanistan in the 1980s, the Salafi militants who returned to their countries maintained transnational links among themselves and formed links with various local Islamist groups. In this way, they created the first vanguard-type cells that Qutb had dreamt of and stimulated the new rise of radical Islamism, now in their own countries. The victory by the Islamic Salvation Front in the first round of general elections in Algeria in December 1991 demonstrated the possibility of Islamists coming to power by peaceful means. The cancellation of the second round of the elections led to the rapid radicalization of Algerian Islamists and prompted them to turn to armed struggle.

[118] Qutb (note 101), p. 231; and Qutb (note 90), pp. 12, 79–80.
[119] Qutb (note 90), pp. 20, 47.

Despite many tactical differences between the followers of the moderate and the more radical currents in modern Islamism, in theory they ultimately seek to advance the same goals. Both the moderates and the radicals seek to spread the Islamist ideology among the masses and to build Islamic states. At least in rhetorical terms, the moderates also ultimately aspire to create, or restore, the supranational, quasi-religious Islamic Caliphate and spread its power to the rest of the world. Both associate socio-economic and political modernization with Westernization and perceive it to be a 'conspiracy against Islam'. For both, religion not only fully dictates the way of life but is also inseparable from the state.

The difference then is primarily in the methods used to achieve these goals and in a different order of priorities. Some Islamist movements stand for building an Islamic state and society by peaceful means only, through persuasion and propaganda. In contrast, the violent Islamists opt for the use of all possible means, including armed struggle, to advance towards the Caliphate.

An inherent political advantage of the Salafi movement, in both its passive fundamentalist and active Islamist forms, is its supra-political character. It is most evident and becomes particularly important in a those Muslim-dominated societies that are split along socio-political, ethnic, clan or any other lines. The all-encompassing nature of the Caliphate as the final goal allows Salafism to bring together groups that otherwise have little in common with one another in political terms. An extremely blurred and distant nature of Salafists' declared ultimate goal is far enough from a concrete political programme to allow very different forces to unite under its banner. The violent Islamists further expedite matters by considering direct participation in jihad to be the main requirement and the shortest way to come closer to the first generations of coverts to Islam.

It is this radical tradition that al-Qaeda, as the core of the broader transnational militant Islamist movement of the late 20th and early 21st centuries, has fed on. In this case the further development of violent Islamism from Qutb to his present followers and interpreters took a very concrete form of personalized succession. Under the strong influence of Qutb and under the deep impression made by his execution, a young Egyptian from a noble family,[120] Ayman al-Zawahiri, the future spiritual mentor and closest associate of bin Laden, founded his own radical Islamist vanguard cell—Islamic Jihad. This group split from the Muslim Brotherhood movement alongside another radical organization, Gamaat al-Islamiya. The Qutbist interpretation of *jahiliyyah* can be clearly traced in all of bin Laden's statements on the West in general and the USA in particular—in bin Laden's words, it is 'the worst civilization witnessed by the history of mankind'.[121] An even more direct ideological influence on bin Laden was provided by the Palestinian-born Islamic scholar-militant Abdullah Azzam, who participated in armed struggle against Israel and in anti-Soviet jihad in Afghanistan. In his time at al-Azhar University in Egypt in the early 1970s, Azzam became acquainted with the Qutb family and al-Zawahiri. He later met with bin Laden when lecturing at King Abdulaziz University in Saudi Arabia and became his ideological mentor. The Soviet intervention in Afghanistan prompted Azzam to revive the 13th–14th century scholar Ahmad ibn Taymiyyah's interpretation of 'the repulsion of the enemy aggressor who assaults the religion and the worldly affairs' as 'the first obligation after Iman' (i.e. after faith itself).[122]

Some moderate Islamist movements have evolved to become more radical and extremist forms and to use violence instead of or, more commonly, in addition to non-violent means. This is the path that Hamas has taken. It developed from a non-violent fundamentalist social movement originating from a Gaza branch of the Muslim Brotherhood network. This branch, which was established well before Israel occupied the Gaza Strip in 1967, suffered repression from Nasser's secular regime in Egypt. For the first two decades of Israeli occupation, the movement devoted itself to religious, social and humanitarian work. It was only in 1987 that it formally established itself as Palestinian resistance movement and joined the armed strug-

---

[120] While in terms of social background, most radical Islamists represented the lower middle class (officers, lower-level officials, clerks, school teachers, traders), some of their leaders and ideologues came from the upper classes or even had aristocratic background.

[121] 'Full text: bin Laden's "letter to America"', *The Observer*, 24 Nov. 2002.

[122] Azzam, A., *Defence of the Muslim Lands: The First Obligation after Iman*, English translation of Arabic text (Religioscope: Fribourg, Feb. 2002), <http://www.religioscope.com/info/doc/jihad/azzam_defence_1_table.htm>, chapter 1. The original text was written in the early 1980s. On Azzam's contribution to the radical interpretation of 'jihad' see section V below.

gle against Israel. Throughout the 1990s and early 2000s Hamas combined violence and terrorism with non-violent protest tactics.[123]

The very possibility of such a transformation does not, however, make the moderate, legalist current that is dominant in modern Islamism less popular or widespread. Nor do radicalization and the resort to violence occur for religious and ideological reasons alone—a decision to switch to violent means may well be dictated by pragmatic social, political and military considerations. Nor is such a transformation into a more radical and militarized organization irreversible. This is exemplified by Hamas, which gradually became a more politicized movement and was capable of winning general elections to the Palestinian Legislative Council in early 2006.

Like many groups of this type, Hamas exists in two dimensions and its goals lie on two levels. At the quasi-religious, ideological level, the movement puts forward fundamentalist goals focused on the ultimate creation of an Islamic state for which 'Allah is its target, the Prophet is its example and the Koran is its constitution'.[124] Ideologically, Islamist groups are not just radical; they aspire to exist in another social, political, religious and ideological dimension, that is, to return to the imagined analogue of the society of the first generations of Muslims. This is a distant goal, which is difficult to achieve and not a concrete political project. While, as they believe, slowly advancing towards that distant goal, Islamist groups such as Hamas have to somehow continue their activities in the meantime. They tend to concentrate their activities on society itself, from the most impoverished sectors of the population to the frustrated parts of the elites.[125] Hamas is a clear example of such a combination of declared religious and ideological goals that have little chance of being realized with far more pragmatic socio-religious and socio-political tasks. The movement's socio-humanitarian work has for years significantly outmatched similar activities by the (secular) Palestinian Authority in terms of their scope, variety and effectiveness. This daily social work with the population and extensive alternative network of socio-

[123] On the origin and evolution of Hamas see e.g. Mishal, S. and Sela, A., *The Palestinian Hamas: Vision, Violence, and Coexistence* (Columbia University Press: New York, 2000), pp. 16–26.

[124] Covenant of the Islamic Resistance Movement [Hamas], 18 Aug. 1988, Article 5, English translation from <http://www.yale.edu/lawweb/avalon/mideast/hamas.htm>.

[125] See Stepanova, *Anti-terrorism and Peace-building During and After Conflict* (note 20), p. 46.

religious relief centres, schools and hospitals has become the main strategic resource of the movement. It helped Hamas gain the support of many Palestinians, especially in the Gaza Strip that brought the movement into the Palestinian Government in 2006.

It is interesting to analyse in more detail the transition to armed violence by Hamas and other groups of this type. The very lack of immediate progress towards their declared ultimate, and unobtainable, quasi-religious goals make these movements particularly dependent on local public support. In contrast to small, marginal political extremist groups in the West, including terrorist groups, these localized Islamist movements cannot exist without popular support and cannot allow themselves to lose this support. It is the vital need for this support that provides a more pragmatic explanation of Islamist groups' extensive social and humanitarian activities. It is also the imperative to keep pace with the prevailing popular mood that often leads them to turn to armed struggle in the first place, as happened to Hamas in the early 1990s.[126] Remarkably, under different conditions the same imperative—to keep pace with the popular mood—may be equally effective in making Islamist pragmatics suspend or halt armed violence, including terrorism. In the early 21st century, armed violence, including terrorism, comprised only a relatively small fraction of Hamas's overall activities. Approximately 90 per cent of these activities continued to be based on social, humanitarian and religious work and, by the middle of the decade, increasingly drifted towards political engagement.[127]

To sum up, in dealing with a relatively large and mass-based movement functioning in a conflict or a post-conflict context, of critical importance is not necessarily whether it employs violence, even if some of this violence takes the form of terrorism. Just as important is whether the movement's ideology and practice can embrace and be integrated with nationalism. If that is the case, then even if a group has employed violent tactics, including terrorism, the option of it joining or returning to the mainstream legalist course of the followers of al-Banna and Maududi is still valid. Its rejection of terrorist means thus remains at least a negotiable scenario in this case.

---

[126] Stepanova, *Anti-terrorism and Peace-building During and After Conflict* (note 20), p. 46.

[127] Council on Foreign Relations, 'Hamas', Backgrounder, 8 June 2007, <http://www.cfr.org/publication/8968>.

However, the ideology of a violent quasi-religious movement (such as the al-Qaeda inspired violent Islamist movement) may be so transnational, even supranational, that it explicitly rejects nationalism. Such an ideology cannot be integrated with nationalism without undergoing a profound change. In that case, the option of such a movement ever abandoning its terrorist tactics is unrealistic. While the gap between ideological rhetoric and practical behaviour may be quite significant for nationalized Islamist actors such as Hamas, for the cells of a supranational post-al-Qaeda movement it is minimal—they are more coherent in matching their actions to their ideology.

In the early 21st century the gravest terrorist threat to international security is not posed by Islamist organizations that effectively combine quasi-religious extremism with nationalism. Nor is it posed by groups that combine violence with a broad range of non-violent functions in their communities and, perhaps most importantly, represent relatively large, territorially-based and mass-based movements. As far as Islamist terrorism is concerned, the main focus of analysis should be on cells and networks functioning in line with the idea of elitist revolutionary Islamist vanguard units composed of the few 'chosen'. It is these cells and networks pursuing an essentially transnational agenda that are most predisposed to emphasize terrorism as their main violent tactic and even as the main form of their activity.

V. Violent Islamism as an ideological basis for terrorism

The Islamic Jihad is a different reality, and has no relationship whatsoever with the modern warfare, neither in respect of the causes of war, nor the obvious manner in which it is conducted.[128]

The victory of the Muslim, which he celebrates and for which he is thankful to God, is not a military victory.[129]

The closest link between radical quasi-religious Islamist ideology and terrorism is provided by extremist interpretations of one of the essential tenets of Islam—the concept of jihad. As noted by one of its earliest and most passionate interpreters, ibn Taymiyyah, jihad 'is a vast

---

[128] Qutb (note 101), p. 227.
[129] Qutb (note 90), p. 124.

subject'.[130] As centuries had been spent on interpretive discussions of the concept by Muslim authors and there is no shortage of recent basic overviews by Western scholars, only introductory remarks are needed here.[131]

According to the moderate interpretations, holy war may take several forms. The principal distinction is between internal (or greater) jihad—religious and spiritual self-perfection and self-purification—and external (or lesser) jihad—armed struggle against aggressors and tyrants. In these interpretations, external jihad is not necessarily the most important, is defensive in nature and is a means of last resort. In contrast, the ideologues of violent Islamism believe armed jihad to be the main weapon in countering the multiple threats and challenges to 'the rule of God' on earth. These threats are posed by forces of secularism (non-believers) and modernization active both from the outside and within the Muslim communities themselves. This extremist view has gained some public following in certain segments of both elites and other social strata of Muslim societies and diasporas. It is supported by the belief in both historical and more recent injustices, ranging from political suppression and direct occupation of Muslim lands to socio-economic marginalization of Muslims by the West. The strongest dissatisfaction is expressed with regard to the policies of the USA, the United Kingdom and Israel. Extremists also build on the lack of legitimacy of the ruling elites and governments in their own countries and have a record of undermining secular nationalist regimes (e.g. in many Arab states).

The distinction between external and internal jihad is not the only one made by moderate Islamic scholars. Another common way to categorize violent ('jihadi') Islamism, which is readily reproduced by Western analysis, is to identify some of its main types, such as liberation, anti-apostate and global jihad.[132] Liberation jihad is armed struggle to drive 'occupiers' and 'non-believers' from the 'native' Muslim lands, be it in Afghanistan, Kashmir, Mindanao or Palestine.

---

[130] Taymiyyah, A. ibn, 'The religious and moral doctrine of jihad', ed. Laqueur (note 104), p. 393. For the complete text in English see <http://www.islamistwatch.org/texts/taymiyyah/moral/moral.html>. On ibn Taymiyyah see also note 110.

[131] For a Western reader, a good basic review of the concept and its historical evolution is provided in the chapter 'Jihad and the struggle for Islam' in Esposito (note 97), pp. 26–70.

[132] See e.g. International Crisis Group (ICG), *Understanding Islamism*, Middle East/North Africa Report no. 37 (ICG: Brussels, 2 Mar. 2005), <http://www.crisisgroup.org/home/index.cfm?id=3301>, p. 14.

This type of jihad is often waged as part of, in combination with or parallel to a broadly nationalist or ethno-separatist insurgency movement that may involve religious or secular groups (e.g. as in the Palestinian territories). Anti-apostate (or internal) jihad targets 'impious' Muslim regimes, for instance in Algeria or Egypt (and is not to be confused with the greater jihad for personal, internal self-perfection). The demarcation between the two is important inasmuch as it is employed by the moderates or even by some of the older generation of Islamic scholars advocating global jihad to distinguish between just and unjust armed struggle. It also helps them rule on the acceptability of civilian deaths, especially among fellow-Muslims when armed struggle takes place within a Muslim country.

These two types of jihad are normally distinguished by the moderates from global jihad. The latter is a transnational (or, more precisely, supranational) movement founded by bin Laden and al-Qaeda with an ultimate goal of establishing Islamic rule worldwide. A series of sub-goals to be achieved along the way includes the support for various liberation and anti-apostate jihads and the global confrontation with the West, especially the USA and its closest allies. As noted in chapter 1, unlike most terrorist actions undertaken by groups waging jihad of the first two types, the use of terrorist means in global jihad qualifies as superterrorism. This categorization is dictated by the unlimited, universalist nature of its ultimate goals and agenda.[133] Thus, if the categorization of jihad into liberation, internal and global is to be accepted, global jihad is the most radical and poses the greatest challenge to international security.

These distinctions are, of course, refuted by the radical Islamic scholars who serve as the main ideologues for the post-al-Qaeda movement. They call for the 'unity of jihad', from the local to the global. In their view, jihad waged against Arab Muslim regimes is legitimate, and there is no restriction on targeting Muslim civilians.[134]

It should be stressed that violent jihad, regardless of its type, level and exact motivations, is by no means a synonym for terrorism and can take different forms and involve different methods and tactics of armed struggle. For instance, a number of Islamist militant groups engaged in fierce fighting in armed conflicts (such as in post-2003 Iraq) do not support indiscriminate attacks against civilians. Issues of

---

[133] See chapter 1 in this volume, section I.
[134] See e.g. al-Libi quoted in Paz (note 89), p. 5.

'legitimacy' of various forms of warfare and methods of armed struggle and of their 'defensive' or 'offensive' nature are regulated by an entire section of Islamic law known as the ethics of jihad (*adb-al-Jihad*).[135] More recently, the so-called law of jihad (*fiqh al-Jihad*) has started to actively develop. At times there may be serious disagreements within the violent Islamist movement on which violent methods are 'legitimate' and which *abd-al-Jihad* is to be followed. A case in point is Algeria since 1992, where sharp disagreements on these issues have become the main driving force behind the major splits and tensions within the violent Islamist opposition.[136]

The most comprehensive and thoroughly developed modern interpretation and justification of jihad as an armed struggle of the second half of the 20th century was put forward by Qutb, who built on all earlier interpreters. It is based on the following basic premises.

1. The goals of jihad are unlimited and universal. They are centred on establishing 'the Sovereignty and Authority of God on earth'. This authority is seen as 'the true system revealed by God for addressing the human life', extermination of 'all the Satanic forces and their ways of life' and abolition of 'the lordship of man over other human beings'.[137] In this radical interpretation, these goals are a logical progression of the unlimited goals of Islam itself.

2. Islam's ultimate goals cannot be achieved without jihad. On the one hand, it is recognized that Islam can resort to methods of 'preaching and persuasion for reforming the ideas and beliefs' while it 'invokes Jihad for eliminating the Jahili order'. Both these tactics are declared to be of 'equal importance'. On the other hand, 'the way of Jihad' is seen as an essential and fundamental requirement for bringing their revolutionary ideas to life.[138]

3. Jihad is interpreted as an active and offensive strategy, rather than being defensive. It is argued that, by viewing jihad as a defensive war only, Muslims deprive their religion of 'its method, which is to abolish all injustice from the earth, to bring people to the worship of God

---

[135] See e.g. the extremist website Electronic Jihad, [Ethics of jihad], <http://www.jehad akmatloob.jeeran.com/fekeh.al-jehad/adab_al-jehad.html> (in Arabic).

[136] International Crisis Group (ICG), *Islamism, Violence and Reform in Algeria: Turning the Page*, Middle East Report no. 29 (ICG: Brussels, 30 July 2004), <http://www.crisisgroup. org/home/index.cfm?id=2884>.

[137] Qutb (note 101), p. 240.

[138] Qutb (note 101), pp. 225–26.

alone'.[139] In this, Qutb apparently draws on teachings by some of his predecessors, particularly Maududi, who considered terms 'offensive' and 'defensive' to be only relevant 'in the context of wars between nations and countries'. They were thus seen as inappropriate for 'an international party' rising 'with a universal faith and ideology' and launching 'an assault on the principles of the opponent' and 'not at all applicable to Islamic jihad'.[140]

4. Armed jihad is interpreted not as a temporary phase, but as a 'natural struggle', 'a perpetual and permanent war' that 'cannot cease until the satanic forces are put to an end and the religion is purified for God in toto'.[141]

5. Finally, the total, all-out nature of jihad is underscored by the rejection of any possibility of a ceasefire, let alone reconciliation, with the *jahiliyyah*. Even if the opponents of Islam consider aggression against it unnecessary, 'Islam cannot declare a "cease-fire" with [the opponents] unless they surrender before the authority of Islam'.[142]

In addition to these core theses, the following, more specific characteristics of jihad as armed violence have crystallized since Qutb's times, particularly in the context of the Palestinian–Israeli conflict, the anti-Soviet jihad in Afghanistan and the 'global jihad' of the late 20th and early 21st centuries.

1. There had been previous challenges to the moderate interpretation of jihad as a collective obligation (*fard kifaya*) of the *umma* that in most cases can be delegated to a few within the Muslim community. However, Azzam's call to reinterpret jihad as an individual obligation (*fard ayn*)—'a compulsory duty on every single Muslim to perform'—marked a critical conceptual shift in modern violent Islamism. An understanding that 'jihad by your person is Fard Ayn upon every Muslim' was central to both al-Qaeda and the post-al-Qaeda transnational violent Islamist movement.[143]

2. An explicit understanding has solidified that armed jihad can be waged against civilians of the 'non-believers' (e.g. Osama bin Laden

[139] Qutb (note 90), p. 56.
[140] Maududi (note 104), p. 400.
[141] Qutb (note 101), pp. 234, 235, 242.
[142] Qutb (note 101), p. 243.
[143] Azzam (note 122), chapter 3.

called in his February 1998 fatwa for the killing of 'the Americans and their allies—civilians and military'[144]).

3. There is no need to abide by certain rules of war that are well-established in Islam and are recognized and emphasized by moderate Islamic scholars and theologists. Among other things, violent Islamists tend to ignore a ban on killing people who are not directly involved in the hostilities, including Muslim civilians and non-combatants.[145]

4. The extremist interpretation of jihad encourages self-sacrifice (suicidal actions) in the course of jihad, extending the centuries-old tradition of martyrdom for faith in Islam to apply to their suicidal tactics, including indiscriminate attacks against civilians.[146]

The followers of the extremist interpretation of jihad that allows for the use of terrorist means are of course very selective in their references to the sacred texts. Much like their opponents, they select only those extracts from the religious texts that justify their 'holy war', often taking them out of context, and tend to ignore those that, for instance, forbid the killing of the innocent. From the Quran, sura 2, verses 190–94 and 216–17, sura 9, verses 5 and 29, and sura 22, verses 39–40, which call for Muslims to fight 'non-believers' in the name of Islam, are some of the most popular and widespread as a religious justification of the use of terrorist means. These selected verses are actively employed by violent Islamist extremists at both the local and global levels. Islamist militants frequently mention, for instance, the call to 'slay them [those who fight you, oppressors, non-believers etc.] wherever ye catch them, and turn them out from where they have Turned you out; for tumult and oppression are worse than slaughter'.[147] Other frequently cited Quranic verses include: 'Fighting is prescribed for you, and ye dislike it. But it is possible that ye dislike a

---

[144] Laden, O. bin, [World Islamic Front for jihad against Jews and crusaders: initial 'fatwa' statement], *al-Quds al-Arabi*, 23 Feb. 1998, p. 3, available at <http://www.library. cornell.edu/colldev/mideast/fatw2.htm> and in English translation at <http://www.pbs.org/ newshour/terrorism/international/fatwa_1998.html>.

[145] See below in this section.

[146] According to the well-established tradition dating back to the Quran, self-sacrifice in the name of God absolves the martyrs from all sins and secures them a privileged place in heaven: 'And if ye are slain, or die, in the way of Allah, forgiveness and mercy from Allah are far better than all they could amass'. Sura 3, verse 157, transl. Yusufali (note 91); see also sura 3, verses 158 and 169.

[147] Sura 2, verse 191, transl. Yusufali (note 91).

thing which is good for you';[148] and 'when the sacred months have passed, slay the idolaters wherever ye find them, and take them (captive), and besiege them, and prepare for them each ambush'.[149]

At the same time the violent extremists tend to ignore, dispute or reject a well-established religious and legal tradition in Islam that forbids the killing of the innocent and emphasizes the defensive nature of jihad. The following are some of the main premises of this tradition.

1. *A general preference for peace over war (against 'non-believers')*. As stated in the Quran, 'if the enemy incline towards peace, do thou (also) incline towards peace, and trust in Allah'.[150]

2. *A general ban on fighting jihad by excessive, or unlawful, means*. This is most clearly articulated in the Quran as: 'fight in the way of Allah with those who fight with you, and do not exceed the limits, surely Allah does not love those who exceed the limits'.[151]

3. *A general ban on the killing of the innocent (regardless of the state of war or peace)*. The Quran equates the killing of one innocent person with the killing of all mankind.[152] It also puts it on a par with another grave offence, polytheism, for which the punishment 'on the day of resurrection' 'shall be doubled'.[153] It is also explicit about the imperative to 'slay not the life which Allah hath made sacred, save in the course of justice'.[154]

4. *The rejection of the killing of Muslims ('believers')*. The Quran threatens anyone who commits such an act with a 'dreadful penalty'. 'If a man kills a believer intentionally, his recompense is Hell, to abide therein (For ever): And the wrath and the curse of Allah are upon him, and a dreadful penalty is prepared for him.'[155]

5. *The centuries-old religious–legal ban on the killing of women and children of the enemy, as well as the elderly, the handicapped and so on*. While some roots of this tradition may be traced back to the

---

[148] Sura 2, verse 216, transl. Yusufali (note 91).

[149] Sura 9, verse 5, transl. Pickthall (note 91).

[150] Sura 8, verse 61, transl. Yusufali (note 91).

[151] Sura 2, verse 190, transl. Shakir (note 91).

[152] Sura 5, verse 32, transl. Pickthall (note 91).

[153] Sura 25, verses 68–69, transl. Shakir (note 91).

[154] Sura 6, verse 151, transl. Pickthall (note 91).

[155] Sura 4, verse 93, transl. Yusufali (note 91).

Quran itself,[156] it is more firmly rooted in the Hadith. According to tradition, when the Prophet Muhammad saw the body of a woman who had been killed, he said 'This is not one with whom fighting should have taken place'.[157] This imperative is repeated in the Prophet Muhammad's statements on several occasions when he is reported as saying 'Do not kill a decrepit old man, or a young infant, or a child, or a woman'.[158] For the earlier fathers of the concept of violent jihad, such as ibn Taymiyyah, who said 'only fight those who fight us', this tradition was still inviolable.[159]

This Islamic religious–legal tradition is so important that some of the most violent Islamist terrorist movements and their ideologues often feel the need to give additional specific explanations of their actions against these categories of civilians. For instance, the Palestinian groups that employ terrorist means insist that all residents of Israel should be treated as potential combatants, as they are allegedly either active servicemen, reservists or are involved in combat support activities. As stated by Yusuf al-Qaradawi, who issued a fatwa on suicidal attacks in the Palestinian context, 'Israeli society is militaristic in nature. Both men and women serve in the army and can be drafted at any moment. . . . if a child or an elderly [person] is killed in such an operation, he is not killed on purpose, but by mistake, and as a result of military necessity. Necessity justifies the forbidden.'[160]

An example of a more general argument is one of the Islamist justifications of the targeting of civilians in the July 2005 London bombings. It was argued that 'the division between civilians and sol-

[156] 'There is no harm in the blind, nor is there any harm in the lame, nor is there any harm in the sick (if they do not go forth [to fight])'. Sura 48, verse 17, transl. Shakir (note 91).

[157] *Sunan Abu-Dawud*, University of South California, Muslim Student Association, Compendium of Muslim Texts, <http://www.usc.edu/dept/MSA/fundamentals/hadithsunnah/abu dawud/>, book 14, no. 2663.

[158] *Sunan Abu-Dawud* (note 157), book 14, no. 2608. The only exceptions to this rule are when women, children and others who are traditionally not expected to take direct part in the hostilities take up arms and thus lose their non-combatant status or when they are so closely intermixed with the armed enemy that they would inevitably fall as 'collateral damage' to the battle.

[159] 'As for those who cannot offer resistance or cannot fight, such as women, children, monks, old people, the blind, handicapped and the like, they shall not be killed, unless they actually fight with words and acts.' Ibn Taymiyyah (note 130), p. 393.

[160] al-Qaradawi, Y., interview in the Egyptian newspaper *Al-Ahram Al-Arabi* (3 Feb. 2001), quoted in Feldner, Y., 'Debating the religious, political and moral legitimacy of suicide bombings, part 1: the debate over religious legitimacy', Inquiry and Analysis Series no. 53, Middle East Media Research Institute (MEMRI), 2 May 2001, <http://memri.org/bin/articles.cgi?Page=archives&Area=ia&ID=IA5301>.

diers is a modern one, and has no basis in Islamic law . . . where every healthy male above 15 years old is a potential soldier'.[161]

Other examples of the early 21st century included attempts to lower the age under which hostages could be regarded as children. Such attempts were made by both the Barayev terrorist group responsible for the seizure of hostages at the Dubrovka theatre in Moscow in October 2002 and, reportedly, by Huchbarov's terrorist group during the September 2004 Beslan school hostage crisis.[162]

6. *A ban on destroying buildings and other property not directly related to an actual battle.*

7. *The inadmissibility of suicidal actions.* This is the interpretation of the Quranic verse 'Nor kill (or destroy) yourselves: for verily Allah hath been to you Most Merciful!'.[163] This is easily superseded in the violent extremists' quasi-religious discourse by the reference to a well-established tradition of martyrdom. In other words, suicidal action is only allowed if it is 'martyrdom' for faith.

There are many disagreements among Muslims themselves, including radical Islamist scholars, on the broader conceptual issues raised by these verses. Whichever interpretation of armed jihad is chosen, in practice the radicalization of Islam among the opposition groups in Muslim-populated regions, especially in conflict areas, often provides these groups with exactly the kind of additional ideological backing needed to use violence. This extends to some of the tactics that may qualify as terrorism—indiscriminate attacks against 'enemy civilians', as well as fellow Muslims (both those perceived as apostates and the innocent). A similar additional ideological backing is often provided

---

[161] [The base of the legitimacy of the London bombings and response to the shameful statement by Abu Basir al-Tartusi], 12 July 2005, transl. and quoted in Paz, R., 'Islamic legitimacy for the London bombings', PRISM Occasional Papers, vol. 3, no. 4 (July 2005), <http://www.e-prism.org/projectsandproducts.html>, p. 5. The original Arabic version is available at <http://www.e-prism.org/>.

[162] Apparently, only those younger than 12 qualified as 'children' for the Barayev group, as only those children were released by the terrorists. E.g. Burban, L. et al., *'Nord-Ost': neo-konchennoe rassledovanie . . . sobytiya, fakty, vyvody* ['Nord-Ost': unfinished investigation . . . events, facts, findings] (Regional Public Organization in Support of the Victims of the 'Nord-Ost' Terrorist Attacks: Moscow, 26 Apr. 2006), <http://www.pravdabeslana.ru/nord ost/sod.htm>, Annex 6. See also e.g. 'Khronika terakta: poslednie novosti!' [Chronicle of the terrorist act: latest news!], ROL, 25 Oct. 2002, <http://www.rol.ru/news/misc/news/02/10/25_017.htm>.

[163] Sura 4, verse 29, transl. Yusufali (note 91).

by the Islamicization of relatively secular—most commonly, broadly nationalist or ethno-separatist—movements.

In post-Baathist Iraq, for instance, it was the Islamicization of the resistance to the US-led occupation that helped the rebels find an ideological, moral and propaganda solution to the increasingly contentious issue of the many civilian Iraqi deaths resulting from their violent attacks. This was not the only way in which the rebels in Iraq were strengthened by the Islamicization of the resistance and the radicalization of Islam among their ranks. Among other things, their appeal to the Salafi religious authorities (*ulema*) in search of moral and legal justification of jihad helped consolidate the resistance movement in 2004–2005 and became the underlying ideological pillar behind the propaganda strategy of its many groups.

In post-invasion Iraq, as well as in a number of other conflict areas, mass casualties among the local population as a result of a combination of anti-state terrorist actions and inter-communal, sectarian or inter-ethnic violence have become daily occurrences. Armed resistance groups have certainly not been the only actors responsible for carrying out such attacks. There have been other perpetrators, including militias affiliated with parties loyal to the foreign presence or even, since 2005, participating in the new Iraqi Government. However, a good deal of such attacks have been blamed on the rebels themselves. In post-2003 Iraq the presence of 'enemy' civilians and civilian objects (the 'natural' targets of terrorist attacks in the context of an ongoing armed conflict) was minimal. It was primarily limited to the employees and property of oil, engineering, communications and other foreign companies, international humanitarian organizations' personnel and diplomats. The overwhelming majority of victims of most forms of violence, including terrorism, were Iraqis themselves.[164] They fall into three broad categories. First are the victims of so-called collateral damage. These are civilians who had been killed or wounded 'by accident', having been caught between the two sides in the course of rebel attacks on military targets and security forces. Second are the frequent intentional victims of terrorist attacks, who are collaborationists of all sorts. They could be the representatives of the government at all levels, including the parties that have joined or support the government. Also, those who try to get employ-

---

[164] Iraq Body Count, 'Year four: simply the worst', Press release, 18 Mar. 2007, <http://www.iraqbodycount.org/analysis/numbers/year-four>.

ment as police or army personnel are often seen by insurgents as collaborating with the occupiers and their proxy regime. Finally, simply the members of a sectarian or ethnic community perceived as relatively loyal to the occupying forces and the new government (especially parts of the Shia community and the Kurds) have been targeted.

In sum, it was the Iraqis themselves that comprised the majority of the victims of both asymmetrical terrorism and symmetrical sectarian and inter-communal strife in Iraq. From the first terrorist attacks on the occupying forces and their Iraqi allies, the armed opposition felt pressured to come up with a convincing ideological justification of the killings of Iraqi civilians and of incurring physical and material damage to civilian objects and infrastructure.

The need for such a justification is by no means equally urgent or pressing for armed non-state actors in all conflict areas. For instance, in the Israeli–Palestinian conflict, the main target of terrorist attacks—the civilian population of Israel (within its pre-1967 borders) and the Israeli settlers on the occupied territories—does not reside thousands of miles away, overseas or on another continent. The enemy civilian targets have been based in the conflict area itself, literally live in the vicinity and have been directly associated by the militant Palestinian groups with their main protagonist—the State of Israel. While the Palestinian terrorist attacks have sometimes resulted in deaths and injuries among Israeli Arabs or Palestinians, the bulk of terrorism victims have been among the 'enemy' civilian population. That fact made the task of political, religious and ideological justification of such actions much easier for the groups responsible.

Another similar example is provided by the struggle for the independence of Algeria (1954–62), which was dominated by secular groups. In the 1950s, the French colonists (some in the third or fourth generation) living in compact, territorially integrated areas comprised up to one million of the Algerian population of nine million.[165] By the autumn of 1955, the anti-colonial resistance movement started to supplement rural and mountain guerrilla warfare tactics with the use of terrorist means in the cities. It was the Algerians of European descent (the so-called *pieds-noirs*) that became the main intentional civilian

---

[165] Galula, D., *Pacification in Algeria, 1956–1958*, new edn (RAND: Santa Monica, Calif., 2006), <http://www.rand.org/pubs/monographs/MG478-1/>, p. xviii. On the *pied-noirs* in Algeria see e.g. Horne, A., *A Savage War of Peace: Algeria 1954–1962* (Macmillan: London, 1977).

targets of terrorist attacks. While hundreds of thousands of Muslim civilians died in the course of the Algerian war of independence, the overwhelming majority of those fatalities were attributed to forms of violence other than terrorism.[166]

In other words, the more accessible, geographically closer and compact is the civilian population of the enemy, the more likely it is to become the main target of terrorist attacks by the local (indigenous) militant non-state actors. For terrorists, violence against the alien or enemy civilians is always easier to justify in the eyes of the community in whose name they claim to act than terrorist attacks that systematically result in casualties among fellow nationals or members of the same population group.[167]

Following the 2003 invasion of Iraq, the extremist interpretation of jihad provided the rebels with a solution to the moral and political dilemma raised by high casualties among the local civilians as a result of terrorist acts. In particular, a call to judge violent actions by their intent, not their actual results was evoked to justify terrorist attacks with mass Iraqi civilian casualties as long as the main target was the enemy. Such a call is typical for the radical interpretations of jihad. According to this approach, 'collateral' or 'casual' victims among the civilian population are seen as acceptable and are justified on condition that the main target was the enemy forces. If the enemy intermingles with civilians, its attempts to use the local population as a cover should not become an insurmountable obstacle to armed jihad. In that case, indiscriminate actions that may result in (mass) civilian casualties among Muslims has still been justified on the grounds that the perpetrators cannot tell the 'innocent' from the 'guilty'. The innocent Muslim victims of indiscriminate terrorist attacks are automatically granted the status of martyr. The only difference with the suicidal militants responsible for killing them is that, while the latter voluntary choose to die in the course of jihad, the former are not consulted on the matter and martyrdom is simply forced on them.

In sum, regardless of the specific justification of terrorist attacks, for Iraqi resistance groups such justification always implied that civil-

---

[166] On casualties in the Algerian war of independence see Clodfelter, M., *Warfare and Armed Conflict: A Statistical Reference to Casualty and Other Figures, 1618–1991* (McFarland: Jefferson, N.C., 1992).

[167] However, even in the course of the Israeli–Palestinian confrontation and during the anti-colonial struggle in Algeria the armed groups felt some need to justify terrorist attacks specifically targeting civilian populations. On such justification see above in this section.

ian casualties were either accidental collateral damage or inevitable losses in a situation where the targeted military and security personnel—foreign or Iraqi—are surrounded by civilians (e.g. during religious ceremonies, festivities, public events etc.).

In contrast, as far as the acts of 'pure' terrorism—attacks that specifically and intentionally target civilian population—are concerned, groups in the Iraqi resistance movement usually have not claimed responsibility for committing them (in a way that can be credibly verified). Among the few exceptions were statements made by the late Abu Musab al-Zarqawi. It is no coincidence that he publicly pledged loyalty to bin Laden and al-Qaeda in October 2004 and merged his militant group with al-Qaeda (renaming it Tanzim al-Qa'idat fi Bilad al-Rafidayn, also known as al-Qaeda in Mesopotamia and al-Qaeda in Iraq).[168] Al-Zarqawi endorsed, for instance, the nearly simultaneous bomb attacks in Baghdad and Karbala in March 2004 at the time of the Shia religious festival of Ashura, which resulted in the death of more than 180 people.[169] His statements could also be interpreted as suggesting his group's responsibility for the attack on the headquarters of one of the main Shia organizations, the Supreme Council of the Islamic Revolution in Iraq, and an attempt on the life of its leader Abdul Aziz al-Hakim in December 2004. Al-Zarqawi's operational doctrine in Iraq combined resistance against occupation with indiscriminate attacks against all 'non-believers' and apostates and an anti-Shia focus.[170] While this method had been approved by a number of younger Islamist clerics, it has faced criticism from some of the radical older ideologues of the post-al-Qaeda global jihad such as Abu Basir al-Tartusi and Abu Muhammad al-Maqdesi.[171]

The Iraqi case also shows that for Islamist groups the justification of attacks against civilians may be greatly facilitated by the blending of terrorism and sectarian violence. The new Iraqi state itself has been formed along sectarian and ethnic lines. It has moved close to becoming a sectarian entity, with some sectarian militias, such as the Kurd-

---

[168] See 'Zarqawi's pledge of allegiance to al-Qaeda: from *Mu'asker al-Battar*, issue 21', transl. J. Pool, *Terrorism Monitor*, vol. 2, no. 24 (16 Dec. 2004), pp. 4–6.

[169] US Department of State, Office of the Coordinator for Counterterrorism, *Country Reports on Terrorism 2004* (US State Department: Washington, DC, Apr. 2005), <http://www.state.gov/s/ct/rls/crt/>, p. 61.

[170] However, even al-Zarqawi refused to take credit for the December 2004 terrorist attacks in the sacred Shia towns of Karbala and Najaf, as well as a number of subsequent attacks of an increasingly sectarian nature. US Department of State (note 169).

[171] For more detail on these debates see Paz (note 89), p. 5; and Paz (note 161), pp. 3, 8.

ish *peshmerga* or the Shia Badr Corps, turning into state-affiliated actors. With the state perceived both as having a strong sectarian bias and as being an agent of the 'occupying forces', the blending of asymmetrical terrorism directed against the state with symmetrical sectarian strife is inevitable. The increasingly sectarian character of terrorism in Iraq has not only made it more deadly. Violence against civilian Muslims has also become easier to justify by emphasizing narrow sectarian or ethno-confessional differences over the more general fellow-Muslim identity. Such justification is further facilitated by emphasizing the links of certain sectarian groups to the 'impure' regime associated with the occupying forces.[172]

While often reinforced by other drivers, as in Iraq, the extremist interpretations of jihad can effectively play the main role in providing specific justifications of armed violence against civilians, including Muslims, whenever there is a need for such ideological justification. The asymmetrical nature of terrorism, whose ultimate target lies beyond its immediate civilian victims, is well understood by Islamist terrorists, their leaders and ideologues. Accordingly, even acts of 'pure' terrorism, intentionally directed against civilians, especially Muslims, may not need any additional or specific justification within the radical interpretation of jihad. These actions can always be interpreted as actions ultimately directed against the main enemy, in one way or another. Of course, there is no need for violent Islamists to go to such extensive lengths in justifying the attacks against enemy civilians. This applies especially to Western civilians, who are believed by violent Islamists to share full responsibility for actions by their democratically elected governments in Afghanistan, Iraq and elsewhere.

VI. Conclusions

Quotations will not suffice, because the perception of the truth relies on the enlightenment of the heart.[173]

The rise of militant Islamism, including Islamist terrorism, at the turn of the 21st century shows the full power of quasi-religious extremism

---

[172] On the blending of terrorism and sectarianism see Stepanova, E., 'Trends in armed conflicts', *SIPRI Yearbook 2008: Armaments, Disarmament and International Security* (Oxford University Press: Oxford, forthcoming 2008).

[173] Azzam (note 122), Final word.

as an ideological basis for terrorism at both the transnational and more localized levels. However, the link between terrorism and religious extremism is not a binding and all-embracing one.

Furthermore, the ideology of militant Islamist groups, including those that employ terrorist means, goes beyond the radical interpretation of the concept of jihad. It is focused on a combination of extremist interpretations of several basic concepts and tenets of Islam.[174] Of these, the basic notion of *imaan* (faith) is perhaps the most important. Jihad, as stressed by radicals from ibn Taymiyyah to Azzam, only comes 'after *imaan*'. The notion of *imaan* is something that provokes scepticism on the part of advocates of the manipulative interpretation of the linkage between religious extremism and terrorism. It is also a stumbling block for those analysts who, in an attempt to rationalize Islamist violence, de-emphasize or disregard the power of religious imperative and conviction for the leaders and rank-and-file members of both local Islamist militant groups and the post-al-Qaeda transnational movement.

Imaan has little to do with theology in the strict sense of the word. It is the power of faith that glorifies acts of violence, including mass-casualty terrorism, for the perpetrators. It is the power of belief that helps explain why for the violent Islamic extremists, the alternative to victory in jihad is not defeat. For militant Islamists, the alternative to victory is either a temporary retreat to consolidate forces (whether it is masked as *hijra* or as a ceasefire), or the ever-present option of dying as a 'martyr'.[175] This distinguishes violent Islamists from their opponents—ranging from moderate Muslims and Muslim regimes to the West—and from secular armed opposition actors.

Among other things, the notion of *imaan* means not only that Islamist terrorists do not accept defeat, but also that they cannot be defeated in principle, at least in their own eyes and in the conventional sense of

---

[174] These, for instance, include extremist interpretation of the basic Islamic concepts of *sabr* (or 'perseverance' in Arabic) which may be summed up as 'Never give up!' and *hijra* (or 'withdrawal' in Arabic). *Hijra* refers to the departure of the Prophet Muhammad from the city of Mecca to Medina in AD 622 (the Hijra). For Islamists, it may mean everything from a complete break with the world of *jahiliyyah* to the possibility of relocation to more secure areas under heavy pressure from a stronger enemy. *Hijra* can also imply temporary suspension of resistance in order to be able to consolidate in exile before continuing jihad with renewed energy.

[175] As noted by ibn Taymiyyah, jihad is generally 'the best voluntary act that man can perform' and anyone who participates in it finds 'either victory and triumph or martyrdom and Paradise'. Ibn Taymiyyah (note 130), pp. 392, 393.

the word. Their ideology allows them to turn even an actual defeat into a spiritual victory, a triumph in the religious sense. As summarized by Qutb, 'When a Muslim embarks upon Jihad and enters the battlefield, he has already won a great encounter of the Jihad'.[176] Furthermore, it is not entirely clear which of two options is more desirable for them. Is it an unrealistic, mythical ultimate victory over the conventionally superior, broadly defined enemy (be it the USA, the West, Muslim regimes corrupted by traumatic modernization or *jahiliyyah* in general)? Or is it an immediate, far more tangible and incomparably more easily achievable death through martyrdom which, they believe, guarantees the shortest and most direct way to God and a distinguished place in heaven?

[176] Qutb (note 101), p. 241.

4. Organizational forms of terrorism at the local and regional levels

I. Introduction: terrorism and organization theory

Identifying forms and constructing models of organizations and organizational behaviour are the main tasks of organization theory. Building on Max Weber's classic organization theory of the turn of the 20th century,[177] organizational studies originally focused primarily on the spheres of business, economics and political economy. Until the 1970s theorists devoted their main attention to the analysis of markets as an organizational form and to their relation with and contrast to hierarchies. Gradually, organization theory expanded its attention beyond economics and started to attract social and political scientists with broader interests.[178] The classification of organizational forms and models was extended to include clans, associations and networks, along with markets and hierarchies. Currently, the main focus of theoretical discussions in the field is on the spread of network forms of organization and on the structural shift from hierarchies to networks.

In the case of terrorism, this general shift towards networked forms of organization is often interpreted as implying a sharp contrast between the 'old' and the 'new' terrorisms. According to this simplistic view, old, pre-11 September 2001 terrorism of ethno-political, leftist and other traditional types is associated with hierarchical models, while the new transnational superterrorism is a synonym for network terrorism.

The analysis of ideological forms of modern terrorism undertaken in chapters 2 and 3 shows that a tendency to draw a sharp line between old and new terrorism has not been very successful, even when applied to the ideological aspects of terrorism in asymmetrical conflict.[179] Over the final decades of the 20th century there was indeed a

---

[177] Weber, M., *The Theory of Social and Economic Organization*, transl. A. M. Henderson and T. Parsons (Free Press: Glencoe, Ill., 1947).

[178] For a good review see e.g. Tsoukas, H. and Knudsen, C. (eds), *The Oxford Handbook of Organization Theory* (Oxford University Press: Oxford, 2005).

[179] See e.g. Lesser et al. (note 77); and Gunaratna, R., *Inside Al Qaeda: Global Network of Terror* (Columbia University Press: New York, 2002).

gradual shift from secular socio-political terrorism to ethno-political and religious or quasi-religious terrorism, or to some combination of the two. Nevertheless, important ideological parallels can be drawn between new transnational terrorist networks and old localized conflict-related terrorism, especially in the case of Islamist terrorism.

It would be more accurate to describe the dynamics of contemporary terrorism not so much in terms of the new–old dichotomy, with the new terrorism sidelining the old, but in terms of ideological and structural developments at different levels of terrorist activity. The most important distinction is thus between terrorism at the transnational (or even global) level and at the more localized levels. The former is a means of struggle that ultimately pursues unlimited goals formulated in accordance with a universalist, globalist ideology. It is not confined by any geographical, national or context-specific limits. Terrorism at the more localized levels is a tactic of asymmetrical confrontation employed by groups and movements that prioritize local, national or, at most, regional agendas.

In the late 20th and early 21st centuries the prevailing type of terrorism practised by organizations with a localized agenda was terrorism by nationalist groups, including ethno-confessional groups. For radical nationalists, a local or, at most, regional context is the most natural level of activity. By definition, radical nationalist groups cannot be universalist or pursue global goals, regardless of the extent of their external links or their additional socio-political or confessional flavour. Meanwhile, at the global level, the main ideology of transnational terrorism of the past—the revolutionary universalism of the radical leftists—has been effectively replaced by quasi-religious violent Islamism as the main ideology of modern superterrorism.

In structural terms, a tendency to view the new network terrorism as a radical departure from the old terrorism of the more traditional hierarchical types is also questionable. Over recent decades the spread of network features has increasingly affected groups at different levels and with varying degrees of centralization and hierarchization. It has produced more hybrid structures that combine elements and features associated with more than one organizational form. Militant groups that employ terrorist means at the local or regional level may also display some new organizational patterns that may not be typical of any of the main known organizational forms (hierarchies, networks, clans etc.).

These structural patterns are reinforced by the rapidly improving and increasingly sophisticated communications capacities of terrorist groups. These upgraded capacities have allowed them to expand their audience and to amplify the demonstrative effect of terrorist attacks (despite the use of otherwise generally standard and not particularly sophisticated technologies, weapons, explosives and other materials). The growing financial autonomy or full financial independence of such groups adds to the complexity of the general picture. A greater degree of financial self-sufficiency has been achieved both by their increasing involvement in criminal activities and by licit means and has been paralleled by the general decrease in state support to terrorism.

At the more localized (i.e. local, national and regional) levels of contemporary terrorism, there are many types of terrorist group, multiple structural models and many patterns combining elements of several organizational forms. It is not the task of this Research Report to produce a thorough review of all these forms and patterns. Rather, this chapter and the next explore whether, in terms of structure, there are any general parallels or sharp contrasts between localized, conflict-related terrorism and transnational terrorism at the global level. A related objective is to assess the impact of the prevailing ideologies of militant non-state actors employing terrorist means on their structural forms and the extent to which their ideologies and structures reinforce each other. It is also important to identify those organizational developments at levels short of the fully transnationalized super-terrorism that best highlight patterns of continuity and change.

II. Emerging networks: before and beyond al-Qaeda

For much of the second half of the 20th century, at least since the 1960s, terrorist means were primarily employed by leftist groups and nationalist (or national liberation) movements. In fact, many of these groups often combined elements of both leftist and nationalist ideologies. The combinations ranged from the prevalence of left-wing or nationalist elements in a group's ideology to the full integration or merger of these elements in the ideology and agenda of a group.

In structural terms, during the last decades of the cold war the most typical type of group to combine conflict-related guerrilla and terrorist activities was a nationalist organization with some degree of left-wing orientation (such as the PLO). A similar widespread combination was a left-wing organization of a nationalist bent (such as the Popular Front for the Liberation of Palestine, PFLP). Such groups, especially radical Marxist or Maoist organizations, tended to have relatively streamlined vertical chains of command and structures that were either fully or significantly centralized.

At a certain point, some of these nationalist left-wing groups started to introduce and increasingly employ network elements, especially at the lower structural levels. An example is the cell-type active service units developed by the IRA from 1977. The purpose of the IRA's structural reorganization was to move away from a strictly hierarchical organization. The hierarchical structure in many ways mirrored the conventional military structure—from the IRA Army Council, via regional brigades and battalions to companies—with a leadership structure at lower levels mirroring that of the higher levels. A set of smaller, more tightly integrated and more autonomous cells— active service units—was introduced to engage in actual attacks, alongside more conventional battalions retained primarily for support activity.[180] In the IRA's case, this structural adjustment was part of a broader shift to a 'long war' strategy. This strategic readjustment was seen as a way out of the stalemate that had resulted from the inability of each side of the armed confrontation to achieve a decisive military success. In this situation, the IRA had to turn to increasingly asymmetrical forms of struggle and patterns of organization. Notably, while the overall organizational shift of the IRA also involved greater emphasis on political and public activity, the introduction of network elements primarily affected and was focused on active militant units carrying out attacks. The introduction of network elements at some lower organizational levels did not, however, radically change the overall hierarchical structure of the IRA and other similar groups and movements and did not turn them into fully fledged hierarchized networks. In sum, hierarchical features and more or less formalized intra-organizational links continued to prevail.

---

[180] The South Armagh regional brigade retained its traditional battalion structure. On the IRA structure and organizational transformation see e.g. O'Brien, B., *Long War: IRA and Sinn Fein 1985 to Today* (Syracuse University Press: Syracuse, N.Y., 1999).

The urban guerrilla: the early network concept

Remarkably, by the time militant–terrorist groups such as the IRA started to integrate the first network elements into their structures, this process not only reflected organic organizational adaptation but already had its own powerful conceptual underpinning. The first modern conceptualization of segmented network resistance through the use of various violent tactics, including terrorism, was made by Carlos Marighella when he formulated his 'urban guerrilla' concept in the late 1960s.[181] Marighella's radical left-wing ideology was an internationalist one. His organizational and tactical recommendations were later widely applied around the world, even though he did not specifically address the international or transnational dimensions of the organizational forms of urban guerrilla warfare. He described a model for the organization of a revolutionary war primarily at the national and regional levels (i.e. in the Brazilian and the broader Latin American contexts).

Marighella was acutely aware of the asymmetrical nature of the armed confrontations fought by insurgents, including those employing terrorist means, and of their enemy's significant, or even absolute, superiority in military force, arms and other resources. The realization of this asymmetry led him to make most of his recommendations in terms of both organizational development and tactics. He perceived a militant non-state actor as doomed to failure if it tried to defend itself against the conventionally superior state on the state's own terms and on its ground, where any non-state actor is weaker by definition. In Marighella's words, 'defensive action means death for us' since 'we are inferior to the enemy'.[182]

Instead, according to Marighella, priority should be given to various innovative types of offensive operations that are not focused on defending a fixed base: 'The paradox is that the urban guerrilla, although weaker, is nevertheless the attacker'.[183] Such a 'technique to attack and retreat', which can 'never be permanent', would be very difficult for the state to counter and could only be effectively carried out by a new type of organization. This organization has to be different from both the centralized hierarchies of many Marxist and Maoist

[181] Marighella (note 30).
[182] Marighella (note 30), p. 16.
[183] Marighella (note 30), p. 16.

political parties and the structural patterns of classic 'rural' guerrillas defending a fixed base. Terrorism was seen by Marighella as just one of several forms of such 'offensive action', but the one that is most asymmetrical in its nature and requires the strongest will and resolve to carry it out. According to him, 'It is an action the urban guerrilla must execute with the greatest cold bloodedness, calmness, and decision'.[184]

The main asymmetrical organizational solution suggested by Marighella is to avoid excessive centralization and hierarchization. This would be a way of denying 'the dictatorship the opportunity to concentrate its forces of repression on the destruction of one tightly organized system operating throughout the country'.[185] This could be achieved through creation of autonomous groups connected to one another and to the 'centre' by shared ideology and direct action rather than through strictly formalized vertical command links. While the centre is still viewed as the main coordinator, the autonomous activity of separate cells—referred to as the 'free initiative'—implies 'mobility, and flexibility, as well as versatility and a command of any situation'. Marighella's urban guerrilla 'cannot let himself . . . wait for orders'.[186]

Conceptually, Marighella managed to go much further in terms of organizational change and adjustment than most of his leftist comrades-in-arms and militant–terrorist groups of other types actually achieved in practice over the next few decades. As early as the late 1960s, his vision of an urban guerrilla movement was already closer to that of a multi-level, hybrid, hierarchized network than to a hierarchical organization employing some network elements (such as the post-1977 IRA).

At the micro level, the urban guerrilla is seen as 'organized in small groups . . . of no more that four or five' (called firing groups). While each guerrilla within a group must be able 'to take care of himself', group cohesion is a critical requirement: 'Within the firing group there must be complete confidence among the comrades'.[187] For Marighella, this requirement was as important as it is now for contemporary cells

---

[184] Marighella (note 30), p. 32.

[185] Marighella (note 30), p. 22.

[186] Marighella (note 30), p. 5.

[187] Marighella (note 30), pp. 11, 13. 'No firing group can remain inactive waiting for orders from above.' Marighella (note 30), p. 14.

of the transnational post-al-Qaeda violent Islamist movement.[188] At the intermediate level, 'A minimum of two firing groups, separated and sealed off from other firing groups, directed and coordinated by one or two persons' makes a firing team. Finally, at the macro level, general tasks are planned by the 'strategic command' and, for dispersed units at lower levels, these tasks take preference. However, the ties linking the strategic command to the rest of the organization should not be too strict or formalized. It is essential to avoid any 'old-type hierarchy, the style of the traditional left' and to avoid 'rigidity in the organization in order to permit the greatest possible initiative on the part of the firing group'.[189] The result is 'an indestructible *network* of firing groups, and of coordinations among them, that functions simply and practically with a general command that also participates in the attacks'.[190]

The parallels between Marighella's urban guerrilla network and contemporary transnational networks guided by a different internationalist and supranational radical Islamist ideology do not end here. According to Marighella, one of the key conditions for such an 'indestructible network' to be effectively coordinated by the strategic command is the extremely general nature and simplicity of its broadly stated goal. An organization should 'exist for no purpose other than pure and simple revolutionary action'.[191]

Perhaps most importantly, Marighella recognized that, to become part of the network, it is not sufficient to share the movement's general ideology. An individual as well as a cell can only become an integral part of the network through direct militant action, including terrorist attack: 'Any single urban guerrilla who wants to establish a firing group and begin action can do so and thus become part of the organization.'[192] Like this purely network method of cell formation, the emphasis on action as the most direct way to join and be accepted by the movement bears a strong resemblance to the way in which cells of the contemporary transnational post-al-Qaeda movement emerge. It is often through direct action that they try to be associated with and be 'legitimized' as part of the broader movement.

---

[188] See chapter 5 in this volume, section IV.

[189] Marighella (note 30), p. 13.

[190] Marighella (note 30), p. 14 (emphasis added).

[191] Marighella (note 30), p. 14.

[192] Marighella (note 30), p. 14.

There are many other striking parallels between this early network vision of the late 1960s and the organizational and tactical dynamics of contemporary, especially transnational, terrorism, with its remarkable spread of network elements. These parallels include an emphasis on the anonymity of action for the network structures,[193] as well as the network cells' ability to adapt to their environment—'to know how to live among the people' and 'be careful not to appear strange and separated from ordinary city life'.[194] There are also references to one of the most effective network fighting techniques that would later become known as the swarming technique. Marighella described it as 'attack on every side with many different armed groups, few in number, each self-contained and operating separately, to disperse the government forces in their pursuit of a thoroughly fragmented organization'.[195]

In sum, it would almost suffice to replace Marighella's notion of an urban guerrilla firing group with a bombing or a suicide-bombing cell for many other organizational and tactical features of his concept to apply to the organizational design of the contemporary violent Islamist movement. In fact, surprisingly few present-day accounts of the transnational violent Islamist movement are as accurate in summarizing some of the main strengths and characteristics of its organizational forms as this early network vision formulated in line with Marighella's concept of the urban guerrilla. This is despite the fact that it dates back some decades before the events of 11 September 2001, was guided by a secular revolutionary ideology and was formulated in a very different context.

---

[193] According to this vision, the network 'method of action eliminates the need for knowing who is carrying out which actions, since there is free initiative and the only important point is to increase substantially the volume of urban guerrilla activity'. Marighella (note 30), p. 14.

[194] Marighella (note 30), p. 6.

[195] Marighella (note 30), p. 22. Swarming is a convergent, breakthrough attack by several autonomous or semi-autonomous, relatively small, dispersed units and cells striking from all direction on the same target. See e.g. Arquilla, J. and Ronfeldt, D., *Swarming and The Future of Conflict*, RAND Documented Briefing (RAND: Santa Monica, Calif., 2000), <http://www.rand.org/pubs/documented_briefings/DB311/>. These authors view swarming as the information-based tactic to apply 'across the entire spectrum of conflicts' and to be employed by state's regular military forces in 'combat operations on land, at sea, and in the air' as much as by the state's opponents (p. iii). More generally, however, the use of the swarming tactic in asymmetrical confrontation against the state appears to be much better tailored for—and to give maximum advantage to—non-state actors. See also chapter 5 in this volume, section II.

These similarities by no means imply that, in structural and tactical terms, nothing new has been introduced by the modern violent Islamists waging 'global jihad' with the use of terrorist means or that little distinguishes them from their secular predecessors. Not surprisingly, most of the major differences are dictated by their differing ideologies—the quasi-religious Islamist universalism of today's post-al-Qaeda terrorist cells and the internationalist secular leftist radicalism of revolutionary groups of Marighella's times. An example is the difference between the indiscriminate nature of attacks by Islamist terrorist cells and Marighella's class-based criterion for target selection. Other examples include the wide use of suicide tactics as opposed to Marighella's emphasis on the need to 'retreat in safety', and the issue of reliance on a broader public movement and mass support.

It is noteworthy that both ideologies that appear to be more favourable to adopting network forms—internationalist left-wing radicalism and modern supranational Islamism—are transnational ideologies. The early network urban guerrilla concept was most popular among, and most actively employed by, internationalist leftist terrorists in Western Europe in the 1970s and 1980s.[196] In contrast, groups and movements in whose ideology nationalism prevailed over internationalist left-wing orientation (such as the PLO) employed many of Marighella's tactical recommendations but showed less interest in his suggested organizational patterns.

Network features and the internationalization of terrorism

Even prior to the 'rise of networks' in the late 20th century,[197] some network elements could be effectively employed by non-state actors of different types and ideological orientations. These actors included militant–terrorist organizations with nationalist, separatist and ethno-confessional motivations and goals that did not go beyond a certain conflict area or local or national context. However, movements that are guided by truly internationalist, even universalist, ideologies

---

[196] Despite this, the spread of network elements in the organization patterns of leftist terrorists in Western Europe was simultaneous with the display of far more hierarchized and centralized models by some of these groups, most typically by Maoists such as the Italian Red Brigades.

[197] Castells, M., *The Information Age: Economy, Society and Culture*, vol. 1, *The Rise of the Network Society*, 2nd edn (Blackwell: Oxford, 2000).

(whether purely ideological or quasi-religious) appear to be most susceptible to the spread of networks. They are better suited for organically developing and operating as network-dominated structures, rather than simply integrating selected network elements into their organizational design.

A related issue is how a group's structural pattern in general, and the extent of integration of its network elements in particular, affects its ability to internationalize its activities. This issue should be addressed against the background of the more general trend towards further internationalization of terrorist activity at the end of the 20th century and in the early 21st century—a process that has taken many forms. This trend is evident, even though the world totals for incidents and casualties for international terrorism, at least for the period since 1998 for which complete data are available, have significantly exceeded by the same indicators for domestic terrorism. In addition, the data show rather uneven dynamics in the internationalization of terrorism, with a number of peaks and troughs (see figures 4.1–4.3).[198] The level of 'internationalization' also varies significantly from one indicator to another and from one type of terrorism to another.[199]

At first glance, it may seem pointless to ask whether it is easier for more strictly structured and heavily centralized groups or organizations with significant network elements and characteristics to internationalize terrorist activity. The immediate answer is apparently in favour of the more networked organizational patterns. However, this question may require a more nuanced answer. It would be more accurate to say that the answer depends on which level of internationalization of a group's activities is being talked about. This level, in turn, is

---

[198] In absolute terms, the main peaks of international terrorist incidents have been in the second half of the 1980s and in the early to mid-2000s (see figure 4.1). International fatality rates also peaked in the late 1980s and early 2000s, but with the latter peak incomparably higher than the former (see figure 4.3). The annual number of injuries in international terrorist incidents has peaked several times since the mid-1990s (see figure 4.2).

[199] Over the last 3 decades of the 20th century, international activity by both left-wing (communist and other leftist) and nationalist groups reached their peaks, in terms of incidents, almost at the same time (during the 1980s; for left-wing terrorism this peak lasted into the early 1990s); while international incidents by religious groups showed first a moderate increase in the mid-1990s and then a major and sharp rise from 1999 until the mid-2000s (see figure 2.1 in chapter 2). Left-wing terrorism consistently caused fewer international fatalities throughout the period (falling to almost nothing since the early 2000s). The first significant peak of international fatalities by nationalist and religious groups dates back to the early 1980s, while the second and far more significant peak can be observed in the first half of the 2000s, especially in the case of religious terrorism (see figure 2.2 in chapter 2).

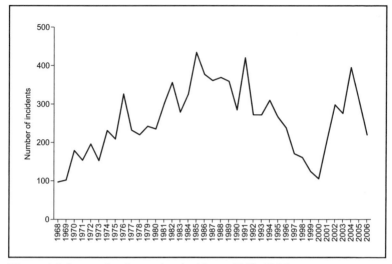

Figure 4.1. International terrorism incidents, 1968–2006
Source: MIPT Terrorism Knowledge Base, <http://www.tkb.org/>.

primarily dictated by the overall level of the group's goals and agenda that are shaped in accordance with its dominant ideology. Separate consolidated localized groups based in different countries or regions may be tied by ideological proximity, such as the solidarity among left-wing nationalist groups or among Islamicized separatists challenging central authorities in their respective countries. If the internationalization in these cases is merely the establishment of contacts between such separate groups, then their relatively centralized and consolidated organizational patterns cannot impede this limited cooperation. Their more streamlined decision-making processes may even aid this cooperation.

It could be argued that the limited internationalization of the terrorist and other activities of radical nationalists of the past decades, especially in the cold war period, was facilitated by the partial embrace of leftist ideology. More recently, a similar role for ethno-separatists in Muslim-populated regions has been played by the growing Islamicization of their ideologies (see below for more detail).

In contrast, superterrorist organizations pursue goals and agendas at a qualitatively higher, transnational, or supranational, level and have a global outlook. It would then be fair to say that ideologically shaped

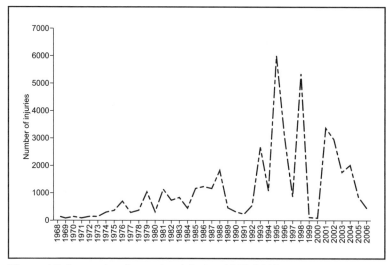

Figure 4.2. International terrorism injuries, 1968–2006
Source: MIPT Terrorism Knowledge Base, <http://www.tkb.org/>.

goals of this type are best suited for organizational patterns dominated by network features,[200] although not necessarily for pure, completely horizontal networks. A typical example may well involve a multi-level network structure that integrates some hierarchical features.[201]

In sum, in shaping a militant group's ability to internationalize its activities, including its ability to effectively carry out international terrorist attacks, organizational patterns are an important, but not decisive, factor. They are of less importance than the overall level of the group's goals and agenda shaped, first and foremost, by its ideology. This dependence closes the circle and further underscores the need to view the two parameters—the ideology and the structure—of non-state actors involved in terrorist activities as interconnected, interdependent and decisively important aspects of terrorism in asymmetrical conflict.

---

[200] This excludes the apocalyptic goals of closed totalitarian religious cults such as Aum Shinrikyo.

[201] In practice, 'pure' horizontal networks are rare and are generally overwhelmed by hybrid structures with a varying degree of network elements and features. See chapter 5 in this volume.

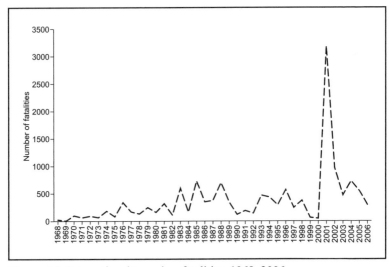

Figure 4.3. International terrorism fatalities, 1968–2006
Source: MIPT Terrorism Knowledge Base, <http://www.tkb.org/>.

III. Organizational patterns of Islamist groups employing terrorism at the local and regional levels

Nowhere is the link between a militant–terrorist group's ideology and its structure more direct than in the case of violent Islamist non-state actors. The structures of these groups and movements in general and their decision-making processes in particular are not very transparent, to say the least. Some general observations can still be made regarding the main types, elements and features of their organizational patterns.

Militant Islamist organizations that are active at a local or regional level are very diverse and the way in which extremist Islamist ideology affects their organizational development varies from one type of group to another. It depends on multiple factors ranging from the group's origin to the way it combines Islamism with other ideologies and motivations (most notably with radical nationalism, including ethno-separatism) and the overall degree of its Islamicization. The latter, for instance, may affect the functions that a group performs, which in turn would be reflected in its organizational structure.[202] On the basis of these criteria, at least four types of organization can be distinguished: (*a*) cross-national Islamist movements that became increasingly nationalist and nationalized (e.g. Hamas in the Palestinian territories and Hezbollah in Lebanon); (*b*) non-nationalist, transnational Islamist movements active in a regional context (e.g. Jemaah Islamiah in South East Asia); (*c*) Islamicized ethno-separatist groups (e.g. in the North Caucasus); and (*d*) Islamicized national liberation groups (e.g. the Iraq insurgency since 2003).

Nationalized Islamists and non-nationalized regional Islamist networks

Both Hamas in the Palestinian territories and Hezbollah in Lebanon emerged as radical transnational Islamic movements. The Sunni Islamist group Hamas grew out of the Gaza branch of the Muslim Brotherhood, that is, it emerged as an autonomous part of a cross-national Islamist network. The radical Shia group Hezbollah (Party of God) emerged in response to Israel's invasion of Lebanon in 1982[203] as a transnationally oriented movement inspired—and sponsored—by Ayatollah Ruhollah Khomeini's revolutionary Iran.[204] The gradual nationalization of these movements has been a long-term process that took decades and has not yet been fully completed. Nor has it necessarily implied a decrease in external support for both movements from Muslim states. The nationalization of these radical Islamic groups has not only been an important development in political and ideological terms but has also had an impact on their organizational evolution.

Both movements perform multiple functions and are engaged in diverse activities. Hamas's initial focus on social and religious work has been supplemented by armed struggle and, increasingly, political

---

[202] The movements formed on the basis of Islamist ideology are, for instance, more likely to be engaged in social and humanitarian work than their secular (e.g. nationalist) counterparts based in the same area.

[203] Israel occupied southern Lebanon in 1982–85 and a smaller border region in 1985–2000.

[204] The origins of Shia Islamism in Lebanon can be traced to the influence of Ayatollah Mohammad Baqir al-Sadr, who founded a revivalist movement in the Shia sacred city of Najaf, Iraq, in the 1960s. Hezbollah derived its ideology from the works of Khomeini and of Musa al-Sadr, the charismatic Iranian cleric who gained a mass following in Lebanon and mysteriously disappeared in 1978. See Hamzeh, A. N., 'Islamism in Lebanon: a guide to the groups', *Middle East Quarterly*, vol. 4, no. 3 (Sep. 1997), pp. 47–54.

activism. In the case of Hezbollah, the original task of armed resistance was later reinforced by socio-religious and political functions. It is thus hardly surprising that the movements' respective structures are quite complex, reflecting their multifaceted (religious, militant, social and political) nature and combining elements of several organizational forms. For instance, the organizational structure of Hamas had displayed many network features since the time when it was a local branch of the Muslim Brotherhood, well before it turned to violence. Hezbollah, which was formed as an armed insurgent movement, originally emerged as a more centralized organization. Its structural model in many ways resembled that of many left-wing national liberation movements of the time, with the main decision-making Consultative Council headed by the secretary-general, supported by the General Convention, Executive Council, Advisory Board and so on. However, it was not a classic hierarchy and actively employed network elements, especially at the movement's lower levels.

The process of nationalization and politicization of the pro-Iranian Shia movement was actively promoted by Hassan Nasrullah after he became Hezbollah's secretary-general in 1992.[205] This process has had a clear impact on the structural development of Hezbollah. It originally emerged as a militant insurgency group guided by imported religio-political radicalism and trying to mirror the ideological and organizational forms of the Iranian model. Gradual ideological and structural transformation has turned it into an increasingly politicized militant movement, with a growing political and social profile. Hezbollah's military organization became a separate and increasingly professionalized component (a quasi-army). Hezbollah has been represented in the Lebanese Parliament since 1992 and the movement has gradually became an essential part of the Lebanese political landscape. It has evolved as a fully fledged multi-level structure based on broad grass roots support, performing basic quasi-state functions for the Lebanese Shia community and being politically active at the national level.

---

[205] In the case of Hezbollah, the term 'Lebanonization' is sometimes used instead of 'nationalization'. On Hezbollah's foundation and evolution see e.g. Hamzeh, A. N., 'Lebanon's Hizbullah: from Islamic revolution to parliamentary accommodation', *Third World Quarterly*, vol. 14, no. 2 (Apr. 1993), pp. 321–37; Ranstorp, M., *Hizb'Allah in Lebanon: The Politics of the Western Hostage Crisis* (St Martin's Press: New York, 1997); and Saad-Ghorayeb, A., *Hizbu'llah: Politics and Religion* (Pluto Press: London, 2002).

For Hamas—an organization that grew out of a set of Islamic social and religious networks—social welfare, humanitarian, educational, religious and other functions continued to amount to a very significant proportion (up to 90 per cent) of the movement's overall activities.[206] The growing nationalization and politicization of the movement required a more streamlined, consolidated and identifiable structure. It has also led Hamas to form an identifiable, collegial political leadership, even though that leadership has remained split between the Palestinian territories and Damascus. The political leadership advances a nationalist agenda, operates on the basis of support provided by the movement's bottom-up socio-religious networks and exercises control over its military branch (the Ezzedeen al-Qassam Brigades). A combination of a strongly nationalist platform with the Islamists' reputation as a relatively incorruptible force and their extensive grass roots social networks is what allowed Hamas to win the Palestinian elections for the first time in 2006.[207]

The nationalization of a cross-national Islamist movement that previously had not tied itself to a nationalist agenda entails a transformation process that may take a variety of forms. It may lead to progressively more active participation in municipal and national elections and the creation of fully legalized and politically well-integrated branches and parliamentary factions. Ultimately, it may even result in the inclusion of an Islamist movement in the national governing structures or its participation in a national power-sharing arrangement as a quasi-state actor.

In other words, resort to nationalism and a significant degree of nationalization play an essential, or even decisive, role in leading radical semi-underground movements to a point where they start acting as political representatives of their ethno-confessional or social communities. Operating in weak, fragile or embryonic states, these movements may fill the vacuum of state power and increasingly and effectively assume some quasi-state functions. Both Hamas and Hezbollah pose as quasi-state actors. They may be ready, if necessary, to join the state and try to transform it from within, as in the case of the

---

[206] Council on Foreign Relations (note 127).

[207] Among other things, the nationalization and politicization of the movement has increased its cross-confessional appeal. In the Jan. 2006 elections, Christian as well as Muslim voters supported Hamas, which also included a Christian candidate on its list. Dalloul, M., 'Christian candidate on Hamas ticket', Aljazeera.net, 25 Jan. 2006, <http://english.aljazeera.net/English/archive/archive?ArchiveId=18115>.

Hamas-led Palestinian Government of March 2006–February 2007 and the Hamas–Fatah 'unity government' of March–June 2007. They can also act as a substitute for the government, claiming to be a more coherent, consolidated, efficient, nation-minded and mass-based force (as in the case of Hamas taking *de facto* control of the Gaza Strip in June 2007).

The quasi-state functions assumed by such non-state actors pose significant political and security challenges in the respective national contexts. While these functions may have controversial effects in terms of a movement's participation in the mainstream political process, they also imply a degree of normalization of its structural forms and its evolution towards more conventional organizational patterns.

These structural developments prompted by and related to the ideological and political evolution of both Hamas and Hezbollah do not yet imply their rejection of armed violence or of an autonomous military role and capabilities. This has been demonstrated both by the continuing militant activity of Hamas after it won the Palestinian parliamentary elections in early 2006 and by the role played by Hezbollah in its asymmetrical armed conflict with Israel in the summer of 2006. What it may help to achieve is a significantly reduced level, or even cessation, of terrorist activity.[208] For Hezbollah in particular, the parallel and interrelated processes of nationalization and politicization have played a decisive role in its turning to forms of violence other than terrorism, ranging from the more traditional guerrilla warfare to the innovative fully fledged asymmetrical confrontation with Israel. In the latter case, a non-state sectarian actor claimed to represent the only genuinely nationalist, effective and efficient military force fighting in the name of the whole of Lebanon and as a substitute for the state due to the latter's supposed ineptness.[209]

---

[208] While Hezbollah was responsible for a series of high-profile terrorist bombings and hostage-taking operations in the 1980s, since its formation it has been primarily engaged in guerrilla warfare against Israeli forces.

[209] Both Hezbollah and Israel insist, although for different reasons, on the ineptness and weakness of the Lebanese state. From Hezbollah's perspective, the sectarian, inefficient and corrupt nature of the Lebanese political system is the main explanation of its inability to defend itself against external enemies, the key disincentive for Hezbollah itself to become fully integrated into this system and the major reason to retain the movement's armed capabilities. From Israel's perspective, the weakness of the central government is the main cause of its inability to prevent the rise of quasi-state actors that pose a significant security risk for Israel.

In the Palestinian case, Hamas was responsible for some of the worst suicide terrorist attacks in the course of the second intifada.[210] However, it restrained its terrorist activity in 2005 and stopped terrorist attacks once it won the parliamentary election in January 2006. While the movement's militants continued to attack Israeli soldiers and launch rocket and mortar attacks against Israel in the summer of 2006 and engaged in violent intra-Palestinian clashes with a rival Fatah movement (e.g. in January 2007), terrorist activity was only conducted by the more radical groups such as Palestinian Islamic Jihad and the al-Aqsa Martyrs Brigades.[211] What is perhaps more important is that the nationalized Sunni Islamists in the Palestinian territories have not been directly associated with the transnational violent Islamist movement inspired by al-Qaeda's example. They have not engaged in any interaction and cooperation with that movement to speak of and generally tend to follow a different organizational, political and tactical path.

The closest example of a militant Islamist movement's evolution in the direction opposite to the processes of nationalization and politicization described above is given by the transnational Jemaah Islamiah network in South East Asia. When the movement emerged in the mid-20th century, it primarily focused on establishing an Islamist state in Indonesia.[212] However, by the end of the century, JI had evolved into a regional network that was no longer tied to any particular territory or any single specific political or national context. Systematic repression by authorities had succeeded in making the JI presence and activities in Indonesia unfeasible for more than a decade. This 'retreat' at the national level played its role in the group's gradual, although highly uneven, transformation into a decentralized regional network since as early as the 1960s. This regionalized movement appears to be unlikely to be transformed, both ideologically and

---

[210] The second intifada refers to the new round of conflict between the Palestinians and Israel that started on 28 Sep. 2000. Examples of suicide attacks include the Mar. 2002 'Passover massacre'. 'Deadly suicide bomb hits Israeli hotel', BBC News, 28 Mar. 2002, <http://news.bbc.co.uk/2/1897522.stm>.

[211] According to the MIPT Terrorism Knowledge Database (note 4), Palestinian Islamic Jihad alone was responsible for 112 attacks in 2006.

[212] On Jemaah Islamiyah see Barton, G., *Jemaah Islamiyah: Radical Islam in Indonesia* (Singapore University Press: Singapore, 2005); and International Crisis Group (ICG), *Indonesia Backgrounder: How the* Jemaah Islamiyah *Terrorist Network Operates*, Asia Report no. 43 (ICG: Brussels, 11 Dec. 2002), <http://www.crisisgroup.org/home/index.cfm?id=1397>.

organizationally, by being drawn into any national political context. Not surprisingly, after the post-al-Qaeda transnational Islamist movement, JI is one of the most networked violent Islamist movements and one of the hardest to deal with.

Islamicized ethno-separatist groups: North Caucasus

The genuine Islamist movements described above are those that were originally formed on the basis of Islamist ideology. In addition to them, attention should also be paid to the structural patterns of groups that emerged as ethno-separatist, radical nationalist or national liberation movements that at first were not associated with religious extremism but later became Islamicized to varying degrees. Groups that always displayed a significant confessional element but were from the start dominated by a nationalist agenda are also of special interest. Islamicized movements of this type are found in Kashmir and Mindanao, but the Islamicized ethno-nationalist resistance in the North Caucasus deserves special attention, primarily due to the strong presence of network characteristics in its structure.

The significant role of network features in the Chechen insurgency, which effectively employed terrorism as one of its violent tactics, is undeniable. However, there has been a tendency in some of the literature to somewhat overestimate either the movement's network character or the degree of its archaization—that is, the extent to which it is dominated by *taip* (clan) structures—or both.[213] Also, in the Chechen and the broader North Caucasian context, attempts to present the insurgency's network features and tactics as an entirely innovative approach are ahistorical. The asymmetrical tactic of fighting against the incomparably superior Russian conventional military forces dates back at least to the Chechen armed resistance to the Russian Empire throughout much of the 19th century. This tactic involved the use of small, tightly knit, dispersed cells that enjoy a great degree of autonomy in both sporadic hit-and-run raids and rudimentary swarming operations.[214]

---

[213] See e.g. Arquilla, J. and Karasik, T., 'Chechnya: a glimpse of future conflict?', *Studies in Conflict and Terrorism*, vol. 22, no. 3 (July–Sep. 1999), pp. 207–29.

[214] On the tactics of the Chechen resistance in the 19th century see the excellent historical account Baddeley, J. F., *The Russian Conquest of the Caucasus* (Longmans, Green, and Co.: London, 1908), pp. 361–64. See also Gammer, M., *The Lone Wolf and the Bear: Three Cen-*

In structural terms, it would be more accurate to describe the Chechen separatist movement of the 1990s and early 2000s as a 'layer cake' of hybrid, hierarchized networks. Along with its many groups, divisions and cells and plenty of semi-autonomous field commanders, it always had an identifiable central command and military–political leadership. As well as some segmented fighting cells, it also had more integrated and consolidated formations, including those that specialized in certain kinds of violent activity. These formations ranged from special reconnaissance or support units to the late Shamil Basayev's own detachment, Riyadh-as-Salihin,[215] which was more specialized in terrorist activity. The movement's network characteristics have been supported and reinforced by elements of clan organization. However, as a whole it cannot be reduced to a form of 'network tribalism' and has evolved as a more advanced structure, especially at its latest stages when it became increasingly regionalized.

The post-Soviet demodernization of Chechnya that resulted from the collapse of the state and the economy and was stimulated and aggravated by the armed conflict itself can best explain parallels between organizational models of post-Soviet rebels and those of the resistance campaigns of the past. However, compared with those campaigns, at the different stages of the post-Soviet rebel movement's evolution, the greatest influence on its organizational formation, transformation and modernization was exerted by two quite disparate factors. At the earlier stage, the command, organizational and battle experience previously gained by some of its founding leaders in the Soviet armed forces and reinforced by the availability of arms stocks left from the Soviet Army played a major role. At the later stages, their influence was succeeded by the growing Islamicization of the movement.

The tactical and strategic thinking and practice of the movement's first generation of commanders and fighters were to a large extent shaped by their military service in the Soviet Army. This experience was not necessarily gained in the most conventional settings (there were veterans of the Soviet intervention in Afghanistan among the rebels). These first generation ethno-nationalists were initially still relatively secularized fighters, many with decades of military experi-

---

turies of Chechen Defiance of Russian Rule (University of Pittsburgh Press: Pittsburgh, Pa., 2006).

[215] 'Riyadh-as-Salihin' means 'gardens of the righteous' in Arabic.

ence. What they brought to the traditional tribal network model of resistance was a higher degree of discipline, coordination and command. They turned the movement into a more hybrid and better organized structure able to go beyond relatively minor hit-and-run attacks against government forces.[216] This may provide a better explanation than that of a combination of clans and networks for some of the separatists' innovative tactics that are not typical of traditional low-scale mountain guerrilla warfare. An example is provided by the rebels' ability to take on massive enemy forces in the course of the 1994–96 First Chechen War, particularly during the 'battle for Grozny' in early 1995 and the Grozny counteroffensive in August 1996.[217]

The increasingly chaotic period of quasi-independence followed the 1996 Khasav-Yurt Agreement that led the Russian Federal Government to temporarily withdraw its forces from Chechnya. While the Islamic revival in the region may be traced back to the late 1980s, since the First Chechen War the radicalization of Islam and the Islamicization of an ethno-separatist insurgency were among the most notable developments in Chechnya—and in the broader region.[218] In organizational terms, its impact went beyond stimulating the internationalization of the movement in general and facilitating the influx of foreign Islamist fighters in particular.

On the one hand, Islamicization has led to further fragmentation, rather than formal consolidation and centralization, of the resistance. It has also resulted in several major splits within the movement. The rise of radical Islam in the movement's ranks made some local warlords concerned about conceding power to Islamists. Instead, these militant actors opted for a mix of Chechen nationalism with a traditional Sufi Islam (such as that of the Qadiriya order). Islamicization was one of the factors that prompted some local armed groups that had joined the separatist insurgency in the First Chechen War to switch sides.

---

[216] See e.g. Kulikov, S. A. and Love, R. R., 'Insurgent groups in Chechnya', *Military Review*, vol. 83, no. 6 (Nov.–Dec. 2003), pp. 21–29.

[217] See Thomas, T. L., 'The battle of Grozny: deadly classroom for urban combat', *Parameters*, vol. 29, no. 2 (summer 1999), pp. 87–102.

[218] See e.g. Malashenko, A., *Islamskie orientiry Severnogo Kavkaza* [Islamic factor in the North Caucasus] (Carnegie Moscow Center/Gendalf: Moscow, 2001); and Tishkov, V., *Chechnya: Life in a War-Torn Society* (University of California Press: Berkeley, Calif., 2004), pp. 164–79.

On the other hand and perhaps more importantly, radicalization along Islamist lines has greatly facilitated the movement's regionalization and significantly reinforced its ability to build cross-ethnic, or even supra-ethnic, networks at the regional rather than the more localized level. These Islamicized networks have emerged as qualitatively different from, and more advanced than, those still permeated by clan and kinship ties and confined to the Chechen ethnic group.[219]

In sum, even for the Chechen insurgency movement, whose structure has traditionally (and exceptionally) been highly networked, further radicalization in the form of Islamicization led to a more networked structure. It has emerged as a qualitatively more advanced structure than the narrow network tribalism and has operated in the regional, inter-, cross- and supra-ethnic contexts. This phenomenon may threaten further transformation of what emerged as the Chechen ethno-separatist movement into a multi-level region-wide set of militant networks—as happened to JI, but in a region far more geographically compact than South East Asia. While armed resistance of this type is more diffuse, it may cause as much trouble in terms of asymmetrical militant and terrorist activity as—and be even more elusive and harder to confront than—a separatist insurgency.

Islamicized national liberation movements: Iraq

In contrast to the separatist movements in the North Caucasus, Kashmir and Mindanao, in post-2003 Iraq anti-occupation and anti-government insurgents have fought for their country to remain a united state and nation. Despite the gradual blending of insurgency with sectarian strife, the fragmentation of violence and the different course and forms that it has taken in different parts of Iraq, the removal of foreign forces from Iraq has continued to be the main goal of the Sunni-dominated resistance and some of the armed Shia groups. While nationalist resistance against the occupation did not in itself lead to coordination of activities by Sunni insurgents and anti-coalition Shia elements, it remained the main characteristic common to both Sunni and Shia radicals.[220] In the four years following the start of the US-led

---

[219] E.g. this is illustrated by the textbook network swarming attack by multiple supra-ethnic Islamicized cells on Nalchik in Oct. 2005.

[220] In 2004 the Shia group Jaysh al-Mahdi (the Mahdi Army) led by Muqtada al-Sadr also fought against the coalition forces. In the mid-2000s insurgent activity by some radical Shia units intensified.

invasion of Iraq in March 2003, most attacks continued to be directed against the coalition, while most casualties were among the Iraqi civilians.[221] The insurgents have also been fighting against the Iraqi Government, which they have seen as being imposed and backed by foreign forces, and against state-affiliated actors of all sectarian identities, but especially the Shia ones.

Since 2003 the Iraqi resistance has become a large-scale, major urban insurgency, a type which is not common for modern armed conflicts. It has been dynamic in terms of its organizational patterns and has developed and changed form almost as fast as the violent Islamist movement at the transnational level has.[222] In structural terms, the Iraqi resistance has not displayed pure network forms. At its earlier stages it manifested itself in separate, uncoordinated, chaotic actions by a number of smaller groups. These first groups, which were primarily motivated by nationalism, did not act as parts of a network, had not yet formed networks and were diverse in their origin. They ranged from remnants of Baathist units to spontaneous protests of groups and individuals that had no Baathist background and emerged 'organically' on the basis of neighbourhood, clan and family, regional and other ties.

By late 2004 the anti-coalition insurgency gradually emerged as a more consolidated set of fewer but larger hybrid, hierarchized network organizations. They combined network characteristics with varying degree of centralization and increasingly resorted to terrorist means, especially suicidal attacks, along with other violent and political tactics, including propaganda.[223] The key role in these organizational and tactical developments was played by the rapid Islamicization of

---

[221] US Department of Defense, *Measuring Stability and Security in Iraq*, Report to Congress (Department of Defense: Washington, DC, Mar. 2007), <http://www.defenselink.mil/home/features/Iraq_Reports/>, pp. 14, 18. For a more detailed explanation of high Iraqi civilian casualties see chapter 3 in this volume, section V.

[222] On the latter see chapter 5 in this volume.

[223] The MIPT Terrorism Knowledge Base (note 4) lists 46 militant–terrorist Sunni insurgent groups—and only 6 Shia groups—in Iraq since 2003 (excluding 2 post-Baath secular groups, some groups that are likely to be criminals masquerading as militants and units that may be part of larger groups). While many of these groups have only claimed responsibility for 1 or 2 terrorist attacks or kidnappings, the largest and most active militant–terrorist groups include Jaish Ansar-al Sunna (Guerrillas of the Army of the Sunna), al-Jaish al-Islami fil-Iraq (Islamic Army in Iraq), and Tanzim al-Qa'idat fi Bilad al-Rafidayn (al-Qaeda in Mesopotamia, which is also known as al-Qaeda in Iraq and by several other names).

the resistance on the basis of radical Sunni Islamism.[224] This gradual ideological consolidation of the resistance movement was based on the merger of militant Islamism and radical nationalism. This process effectively blurred the ideological and structural differences between the foreign Islamist fighters with transnational connections and Iraqi Islamist nationalist Sunni groups engaged in anti-coalition violence and, increasingly, in sectarian strife.

This convergence between radical Islamism and nationalism had dramatic impact on the insurgents' resolve and tactics, including growing emphasis on terrorism and suicidal attacks.[225] Tanzim al-Qa'idat fi Bilad al-Rafidayn (al-Qaeda in Iraq) has remained one of the larger groups of the resistance even after Abu Musab al-Zarqawi's death in 2006 and has continued to use foreign fighters, although the majority of the group's members are Iraqis.[226] In early 2006 this group formed the core of the Mujahideen Shura Council—an umbrella coalition that later declared the 'the foundation of the righteous state, the Islamic state' in Iraq based on sharia.[227]

Although the influence of transnational terrorist networks on the dynamics of violence in Iraq appears to be exaggerated, the rise of violent Islamism and its convergence with nationalism in Iraq has played a role of a broader international significance. Since the attacks of 11 September 2001, the Islamicized Iraqi 'national liberation' resistance has become perhaps the most powerful political and quasi-religious symbol for transnational violent Islamism. It has provided a powerful motivational impulse and mobilizing influence for the existing and new cells of the post-al-Qaeda violent Islamist movement operating in different parts of the world and guided by a global vision.

---

[224] For an analysis of this transformation see International Crisis Group (ICG), *In Their Own Word: Reading the Iraqi Insurgency*, Middle East Report no. 50 (ICG: Brussels, 15 Feb. 2006), <http://www.crisisgroup.org/home/index.cfm?id=3953>.

[225] See chapter 3 in this volume, section V.

[226] This fact is recognized by the US Government, as well as the fact that foreign militants in general made up just 4–10% of the approximately 20 000 rebels in Iraq in 2006. US Department of State, Office of the Coordinator for Counterterrorism, *Country Reports on Terrorism 2005* (US State Department: Washington, DC, Apr. 2006), <http://www.state.gov/s/ct/rls/crt/>, p. 131. The US bipartisan Iraq Study Group report estimated the number of foreign 'jihadists' in Iraq in 2006 at 1300. Baker, J. A. III and Hamilton, L. H. (co-chairs), *The Iraq Study Group Report* (Iraq Study Group: 2006), <http://www.bakerinstitute.org/Publication_List.cfm>, p. 10.

[227] See MIPT Terrorism Knowledge Base (note 4); and Mujahideen Shura Council in Iraq, 'The announcement of the establishment of the Islamic State of Iraq', 15 Oct. 2006.

The only major factor that has acted against the Islamist–nationalist convergence along the lines described above has been the further fragmentation of violence in Iraq and especially the rise of intra-sectarian (including intra-Sunni) tensions in 2006–2007. US governmental sources have consistently tended to exaggerate the differences between 'Iraqi insurgents' and 'foreign mujahideen'.[228] Both the coalition powers and the Iraqi Government have done their best to encourage any divisions between Sunni tribal groups and the more radical al-Qaeda in Iraq and other strongly Islamist groups.[229]

However, the divide-and-rule approach also has major negative security repercussions contributing to further instability and fragmentation of violence on the ground. It cannot provide a lasting solution to the problem. A more constructive and fundamental way to weaken both the ideological link and organizational ties between the Iraqi insurgency and transnational Islamism would be to promote an ideology that is at least as powerful and appealing at the national level as Islamism. It would imply supporting and encouraging Iraqi cross-sectarian Arab nationalism, instead of suppressing it. In other words, the optimal strategy would have been almost exactly the opposite to the one that has been followed by the international interveners in Iraq and which has contributed to both sectarianism and inter-ethnic tensions. Iraqi Arab nationalism, even in its radical forms, appears to be the only ideology that can unite Iraqi's main Sunni and Shia communities. It is the main force that could keep the country together and act as a counterbalance to both transnational Islamism in Iraq and the symbolic meaning of Iraq for violent Islamist cells throughout the world. Only genuine home-grown groups and movements that come closest to Iraqi Arab nationalism and may have some cross-sectarian appeal could form the basis for meeting this challenge. While these forces, such as the Muqtada al-Sadr's Mahdi Army, may be very radical and non-secular and are strongly defiant of the foreign occupation, they may remain the only ones who have not discredited their nationalist credentials.

---

[228] See e.g. US Department of State (note 226), p. 130.

[229] See e.g. Knights, M., 'Struggle for control: the uncertain future of Iraq's Sunni Arabs', *Jane's Intelligence Review*, vol. 19, no. 1 (Jan. 2007), pp. 18–23. See also Stepanova (note 172).

IV. Conclusions

Attempts to draw a precise dividing line between 'new', post-11 September 2001, loosely organized, transnational network terrorism and 'old' terrorism of the more traditional organizational types have not been conclusive. This approach views the new network forms of terrorism as a radical departure from the old localized and hierarchized forms, as if the latter had not been structurally evolving over recent decades. Instead, it appears that network elements and features have been and are increasingly employed by terrorist groups of different types at all levels from the local to the global. At least in this sense, an organizational difference between terrorism at the transnational and the more localized levels may be more gradual than substantial.

Similarly, strict hierarchical forms can manifest themselves in both the structures of groups with a more localized agenda and, for instance, the organizational patterns displayed by superterrorist apocalyptic religious sects with a universalist agenda. Naturally, more centralized and hierarchical forms are more widespread at the localized levels.

Nationalism—whether of ethnic, ethno-confessional or a broader, civic or cross-ethnic or cross-confessional type—is the strongest force that can tie a militant organization to a certain location, territory or national context and streamline and conventionalize its structure. The closer that such an organization is tied to a territory and to a localized context, the stronger is the pressures on it to take on quasi-governance functions and the more it ultimately sees itself as a new, revised or better analogue of the state it is fighting. The more that such a group sees itself in that way and structures itself in line with this vision, the easier it is to identify, deal with and transform. This is particularly important when dealing with mass-based and popular Islamicized and Islamist armed movements that may employ terrorism as one of their tactics but cannot be defeated by conventional military means. The more that such a movement is nationalized and immersed in a national political context, the more realistic are the chances that its militant hardliners will gradually become marginalized. The movement may also be more willing to reject the militant tactics that are most deadly for the civilian population. Finally, any links of such a nationalist–

Islamist movements with transnational violent Islamism of the post-al-Qaeda type are more likely to erode.

However, the growing organizational divisions between the radicals and the more moderate, pragmatic and nationally focused forces within such a movement are not sufficient to effectively co-opt its more pragmatic leaders and forces into the mainstream political system. These efforts can only succeed if they are integrated into the broader process of transformation of the state that these Islamist nationalists have been fighting against. In contexts as different as post-2003 Iraq and Lebanon, the functionality or legitimacy of the state itself and its deeply divisive and sectarian character both discourages the violent Islamist non-state actors to associate with it and allows them to claim to pose as the more genuine and nationally oriented forces.

5. Organizational forms of the violent Islamist movement at the transnational level

I. Introduction

It has become a commonplace to refer to the general spread and evolution of network structures and, more specifically, to their employment by and impact on anti-system actors such as terrorist groups. In particular, since 11 September 2001, al-Qaeda and the broader post-al-Qaeda transnational violent Islamist movement have often been described as 'model' or 'pure' terrorist networks. They have been commonly viewed as a radical departure from the 'old' terrorism practised by groups of the more traditional hierarchical type.[230]

As time continues to pass since 11 September 2001, this simplified interpretation is becoming ever more superficial. In particular, it is no longer sufficient to refer to al-Qaeda and the post-al-Qaeda movement as standard networks, described in the most general terms. The problem is not simply that analysts find it hard to keep pace with the rapid changes in organizational forms of the transnational violent Islamist movement. Years after the dramatic terrorist attacks of September 2001, the time for simplistic explanations is over. There is a pressing need for a more nuanced approach by experts in organization theory in their study of the structures of underground terrorist and other anti-system actors. It is also required from analysts specializing in various aspects of political (ideological, religious and quasi-religious) extremism and violence, including terrorism, and other challenges to national and international security. These challenges are commonly known as unconventional or non-traditional threats, but it is more accurate to refer to them as recently securitized threats, given that they are no longer peripheral issues.

Much has been said and written about the network characteristics that allow transnational actors—ranging from socio-political activist associations to militant movements that employ terrorist means—to

---

[230] See Gunaratna (note 179), pp. 54–58, 95–101 etc.

'think globally, act locally'.[231] But which of these network features are most typical of the transnational violent Islamist movement? In what way does its organization resemble other standard modern social networks and what makes it different? In which direction do its network organizational forms and elements evolve? How do they interact and integrate with elements and features of other organizational forms within the movement's structural framework? What impact does the movement's ideology have on its structural patterns? What other factors have an impact on its organizational development? These are some of the questions about the structures of contemporary transnational terrorism—and especially the post-al-Qaeda movement as its most advanced and dynamic form—addressed in this chapter.

II. Transnational networks and hybrids: combinations and disparities

Analysis of the structural patterns of modern terrorism, especially of its transnational forms, has been dominated by 'organizational network' theory. According to this theory, a network is a specific, separate organizational form that has gained force in an age of rapid development of information and communication technologies.[232] In this information age, network structures appear to have some important advantages over other organizational forms. For instance, compared to hierarchical structures, network organizations are more flexible, more mobile, better adapt to changing circumstances and are more stable during system shocks and at times of crisis. For a certain structure to function as a network it is not sufficient for its main elements to be linked by horizontal ties (as opposed to the prevalence of vertical ties in hierarchies). For it to be a network, all of its elements must both view themselves as parts of a broader network and be ready to act as a network. From the organization theory perspective, the

---

[231] This is a slogan of Friends of the Earth, an international environmentalist movement founded in the USA in 1969 and structured as a network of autonomous grass roots groups. Authorship of the slogan is disputed and attributed to several people, including the network's founder, David Brower.

[232] Castells (note 197); Arquilla, J. and Ronfeldt, D. (eds), *Networks and Netwars: The Future of Terror, Crime, and Militancy* (RAND: Santa Monica, Calif., 2001), <http://www.rand.org/pubs/monograph_reports/MR1382/>; and Arquilla, J. and Ronfeldt, D. F., 'Netwar revisited: the fight for the future continues', *Low Intensity Conflict & Law Enforcement*, vol. 11, nos 2–3 (winter 2002), pp. 178–89.

main characteristic of any network is its non-hierarchical, decentralized character, which explains the primary focus of this theory on the conflicts, correlations and interactions between networks and hierarchies.

In contrast to organizational network theory, 'social network' theory explores all sorts of interlinkages between social actors and the social structures that stem from and are based on these interlinkages.[233] Rather than viewing the network as a specific, separate organizational form, this theory views it as a system of interrelations in society that characterize all forms of social life. For social network theorists, a more general distinction between an informal network and a formal organization is more important than the contrast between network and hierarchical organizational forms.[234] Any organization, especially a relatively large one, even if it is decentralized to a significant extent, requires at least a minimal set of hierarchical features. In contrast, a network in principle lacks a central leadership presiding over a strict hierarchy. While the elements of a network are interconnected, they are autonomous and are not subject to direct, formal orders 'from above'.

General trends in the development of networks

While keeping in mind these two broad theoretical approaches, it is useful to consider the four broadly acknowledged general trends in the development of the network characteristics of modern non-state organizations, including anti-system actors such as terrorist groups.

The first trend is the general spread of network forms, especially among non-state actors. Groups that display the key network features gain considerable advantages in asymmetrical confrontation against the less flexible and less mobile state structures. The lack of a strict hierarchy and of a single structured central leadership exercising direct control over subordinate units complicates the task of destroying these movements. The spread of network features can be traced in the organizational development of groups of different types, goals and orientation ranging from criminal or militant–terrorist armed anti-

---

[233] See e.g. Scott, J., *Social Network Analysis: A Handbook*, 2nd edn (Sage: London, 2000).

[234] See e.g. Nohria, N. and Eccles, R. G. (eds), *Networks and Organizations: Structure, Form, and Action* (Harvard Business School Press: Boston, Mass., 1992).

system actors to environmental and civil society groups and movements. Transnational non-state activist networks and associations range from anti-globalists to grass roots movements against the use of landmines or against 'blood diamonds'.[235] These are standard examples of modern networks that are actively challenging states or trying to engage them in addressing the movement's main issues of concern. These examples may in fact be far more typical for networks as an organizational form—and are certainly far more transparent—than the violent transnational Islamist movement inspired by al-Qaeda whose structure is more than just a standard network and is much more difficult to study. More generally, excessive attention to the use of network forms of organization by terrorist, criminal and other underground structures presents a somewhat distorted picture. It underestimates the positive potential of network structures in the information age.[236]

Second, whichever theoretical approach is applied, in practice neither the contrast between networks and hierarchies nor the distinctions between informal decentralized networks and formal organizations are strict dichotomies. Nor do they adequately reflect the complex dialectic nature of modern organizational models, most of which are mixed, hybrid structures. The basic distinctions between networks and hierarchies do not mean that there is no space for a broad range of intermediate structures. In the spectrum of structural models, most organizations—including terrorist groups—fit somewhere between the two extremes of a pure network and a pure hierarchy. Most display both network and hierarchical elements, sometimes in combination with elements of other organizational forms, such as clans. In a dynamic process of organizational development, this combination

---

[235] The anti-globalization movement (also known as the Global Justice movement) is an umbrella term for a number of social movements that oppose some of the controversial aspects of globalization, which is seen as deepening or even generating social injustice and inequality, such as 'corporate globalization', free-trade agreements etc. From 1999 to mid-2007, the anti-globalization movement has organized up to 50 large-scale transnational actions, mostly at the time of large international summits. The International Campaign to Ban Land-Mines is a network of more than 1400 NGOs in 90 countries. See <http://www.icbl.org/>. The movement against 'blood diamonds' led to the establishment of the Kimberley Process Certification Scheme in Nov. 2002, setting up an internationally recognized certification system for rough diamonds and national import–export standards adopted by 52 governments.

[236] An attempt to challenge this simplistic view and to highlight both positive and negative implications of the 'rise of networks' among non-state actors was one of the central themes of the landmark study eds Arquilla and Ronfeldt (note 232).

may change from, for example, a relatively loose organization balancing hierarchical and network features, such as in al-Qaeda, to a more decentralized movement. This more decentralized organization pattern employed by the post-al-Qaeda transnational violent Islamist movement retains multi-level coordination and some informal vertical ties but is dominated by network forms.

Third, with all the attention that has been paid to the network characteristics of the modern superterrorist networks (i.e. primarily the transnational violent Islamist movement), it would be a mistake to say that network models are found only in the relatively recent phenomenon of superterrorism. As discussed in chapter 4, some basic network characteristics are also to be found in more traditional types of terrorist group. To a certain—and growing—extent, these characteristics have been an essential part of organizational design for a number of groups that were engaged in violent activity at a more localized level. These groups' agendas have not gone beyond a national framework or a particular armed conflict. Examples range from the IRA in Northern Ireland and Sendero Luminoso in Peru to the Islamist Hamas or the more secularized al-Aqsa Martyrs Brigades (Arafat Brigades) in the Palestinian territories.

In recent years, analysts as well as practitioners have paid much attention to 'the rise of networks' in general, and the spread of network structures among anti-system actors in particular. However, they have often forgotten that the first attempts to conceptualize segmented network urban guerrilla and terrorism structures and tactics date back to the late 1960s.[237] In addition, many of the tactics typical of modern network warfare, such as swarming, are no less popular among the localized militant groups combining guerrilla and terrorist means than among the cells of the post-al-Qaeda movement.[238]

While network forms prevail in the structural models of superterrorism, they do not do so absolutely. The Japanese cult Aum Shinrikyo, which fully qualifies as a superterrorist group due to the global nature of its goals and agenda and its readiness to use unlimited means to achieve those goals, was structured as a strict vertical hierarchy.[239]

---

[237] See chapter 4 in this volume.

[238] On swarming see note 195.

[239] Aum Shinrikyo launched 17 attacks using chemical or biological weapons. In the most deadly of these, on 20 Mar. 1995 the chemical nerve agent sarin was released on Tokyo underground trains, killing 12 people and injuring more than 1000. Monterey Institute of

In this context, it is worth recalling that the main defining criterion of a superterrorist group is not its network structures (as opposed to the more hierarchical organizational forms of the traditional types of terrorism), but the level and scope of its goals and agenda. Of critical importance is whether these goals are global (and unlimited) or are limited to a more localized context.

The spread of network elements gives tangible comparative advantages to terrorist groups at all levels. If there is any major difference between the more traditional terrorism at the local and regional levels and superterrorism in terms of organization, it is in the varying degrees of correlation of network and hierarchical elements. Naturally, for a transnational violent Islamist movement with a virtually global outreach and unlimited goals, the role played by network characteristics is much higher than it is for a more localized group. As this chapter shows, any more substantial disparities in the way these groups function cannot be explained in terms of organizational forms alone—factors of an ideological and social nature need to be drawn in.

Fourth, the issue of concern is not just the network character of the transnational violent Islamist movement. Its organizational patterns go beyond those of a standard modern anti-system network that, for instance, characterizes the anti-globalist movement. The advantages given by the standard network features in asymmetrical confrontation against the less flexible and less mobile state structures are detailed above. Nonetheless, 'classic' networks also have serious drawbacks and weaknesses. First and foremost is the difficulties they can experience when faced with the need to make strategic political–military decisions and to put them into effect. They also lack purely organizational mechanisms to ensure that these decisions are followed by all the main elements within the network and to exercise control over the implementation process. The informal and ulterior nature of the links between various network elements allows such an organizational system to function effectively only under certain conditions.[240] The mere fact that multiple cells form a network and even their basic ideological proximity may not suffice to impose upon them strong and stable mutual obligations to engage in violent activity, especially in the form of terrorism against civilians.

International Studies, Center for Non-Proliferation Studies (CNS), 'Chronology of Aum Shinrikyo's CBW activities', 2001, <http://cns.miis.edu/pubs/reports/aum_chrn.htm>.

[240] In specialized literature these informal links are commonly referred to as 'latent' links.

In sum, modern transnational terrorist networks such as the post-al-Qaeda movement display an amorphous, multi-layered structure and loose and ulterior links between different elements. A lack of a strict vertical chain of command and informal leadership patterns at the macro level is coupled with multiple and diverse cell patterns displaying varying combinations of network and hierarchic features at the micro level. The main question then is why, despite all these characteristics, this movement manages to act effectively and seems to function as one organism. How does a structural model that displays the main network characteristics—even if they are combined with elements of other organizational forms—manage to effectively neutralize its inherent weaknesses?

III. Beyond network tribalism

Functional–ideological networks

According to organizational network theory, the structural development of al-Qaeda into the broader, more fragmented and dispersed post-al-Qaeda movement displays a transitional organizational pattern. It has evolved from a more formalized organization to a more amorphous, decentralized network of cells that spread and multiply in a way that, in terms of organizational form, closely resembles franchise business schemes. These cells share the movement's transnational violent Islamist ideology, follow general strategic guidelines formulated by its leaders and ideologues and use the name of 'al-Qaeda' as a 'brand' but are not necessarily formally linked to it in structural terms.

This creeping network displays at least some of the main characteristics of a segmented polycentric ideologically integrated network (a SPIN structure)—one of the most advanced types of network described and studied to date.[241] The segmented nature of a SPIN

---

[241] The concept of a SPIN structure was formulated by anthropologist Luther Gerlach and sociologist Virginia Hine in the early 1970s on the basis of their studies of civil rights groups and social protest movements in the USA in the 1960s and early 1970s. See Gerlach, L. P. and Hine, V. H., *People, Power, Change: Movements of Social Transformation* (Bobbs-Merril: New York, 1970); Gerlach, L., 'Protest movements and the construction of risk', eds B. B. Johnson and V. T. Covello, *The Social and Cultural Construction of Risk: Essays on Risk Selection and Perception* (D. Reidel: Boston, Mass., 1987), pp. 103–45; and Gerlach, L. P., 'The structure of social movements: environmental activism and its opponents', eds Arquilla and Ronfeldt (note 232), pp. 289–310.

structure means that it is made of many cells. Its polycentric character implies that it lacks single central leadership, but has several leaders and central nodes. Its network structure indicates that its various segments, leaders and central nodes are integrated into a network by means of structural, ideological and personal links. SPIN structures demonstrate a very high level of structural flexibility and adaptability. The model allows, for instance, social protest movements to effectively resist suppressive measures by states, to penetrate all strata of society, and to promptly and effectively adapt to the rapidly changing political and social environment.

The main integrating force for a network that approximates the SPIN structure is its shared ideology. To emphasize this connection, in this Research Report the term used to refer to most networks of this type—including both violent and non-violent activist movements—is 'functional–ideological network'. Using modern means of communication, shared ideology helps connect the fragmented, dispersed, isolated or informally interlinked elements of modern networks. This organizational form dominates many social protest movements in the West, as well as some broader campaigns such as the anti-globalist movement. As noted above, for modern functional–ideological networks, common ideological beliefs and values play an even higher role as the main connecting and binding principle than they do for more traditional types of anti-system group.[242]

The post-al-Qaeda movement is often seen as a network embodiment of the ideology of 'global Salafi jihad'. However, even some of the strongest advocates of this view have come to understand that the transnational violent Islamist movement cannot be reduced to a standard impersonalized functional–ideological network. The transnational violent Islamist movement's underground cells emerge in different political contexts and are dispersed in many parts of the world. If they are tied into a broader decentralized network, it is through some informal, hidden links. These characteristics do not appear to match the active, effective and seemingly well-coordinated manner in which these cells carry out their terrorist activities. Indeed, the scope and level of the post-al-Qaeda movement's operational activities require a much higher level of intra-organizational coherence and trust than can be provided by religious and ideological beliefs and goals alone, especially if the latter are formulated in a very

[242] See e.g. chapter 1 in this volume, section III.

general way. Against this background, some analysts have started to doubt whether ideological, including religious and quasi-religious, goals and beliefs suffice to explain how the transnational violent Islamist movement functions so effectively, at least at the micro level of individual cells. This underscores the need to supplement the ideology-centred perspective with more nuanced approaches. The approach that focuses excessively on militant Islamism as the single driving and organizing force of the transnational violent Islamist movement needs to be corrected and adjusted, if not radically revised.

Network tribalism: a critique

One way to revise the approach centred on functional–ideological networks is based on the following assumption. It argues that the lack of a single central leadership and the multiplicity of real and 'virtual' leaders that is typical of many modern transnational networks forces the network elements to resort to various consultative and consensus-building mechanisms in the decision-making process. Such mechanisms were typical for many pre-hierarchical clan and tribal organizational forms and social systems. From that, some analysts immediately—and somewhat hastily—concluded that the post-al-Qaeda movement points to the revival of elements of tribalism at a new, network level. In other words, it provides an example of the integration of modern post-hierarchical elements into a structure that is closer to archaic, pre-hierarchical forms. This approach could be traced in the evolution of the views of one of the leading network theorists, David Ronfeldt. He turned from an interpretation of al-Qaeda as a super-modern transnational network to a description of a network of al-Qaeda-affiliated groups as a semi-archaic 'global clan waging segmental warfare'.[243] According to this network tribalism approach, the transnational violent Islamist movement is both a reaction to the information revolution and other aspects of globalization and a force that makes full use of the achievements of the information age in order to revive aggressive clan-based tribalism on a global scale.

In contrast to hierarchies, networks and markets, the clan form of organization is based on the family or broader kin relationships, both nuclear and linear. They are usually reinforced by the idea of a

---

[243] Ronfeldt, D., 'Al Qaeda and its affiliates: a global tribe waging segmental warfare?', *First Monday*, vol. 10, no. 3 (Mar. 2005), <http://firstmonday.org/issues/issue10_3/>.

common origin often traced back to some mythological ancestor. In terms of structure, clans are egalitarian, segmented entities that have no power-based leaders in the hierarchical sense and no strict vertical links of subordination. Everything is decided by consensus through consultation upon advice from the most respected and experienced clan members (usually, the 'elders'). For clans, the prevailing mood is that of collective responsibility and intra-clan solidarity, which does not, however, extend to those who are not clan members. Tensions and conflicts are resolved by means of compensation or revenge. The main goal and value for the clan is not so much power (as in hierarchies) or profit (as in markets) as honour and respect by other members of the clan.[244]

The network tribalism concept insists that individual cells of the transnational violent Islamist movement are not built as impersonalized network elements. Rather, they are created on the basis of family, kinship and clan ties and form what at first sight may resemble a traditional extended family. From the point of view of both organizational network and social network theories, clans and networks do indeed have something in common—the absence of a formally institutionalized hierarchy. Clan and network features may thus overlap to some extent. However, clans and networks are not identical and are not driven by exactly the same dynamics.

According to Ronfeldt, the focus on the clan model is more adequate than the emphasis on the network paradigm. He points to such inherent clan characteristics as infinite loyalty to one's own clan, sharp distinctions made between the notions of 'them' and 'us' and revenge as a 'natural' form of violence. He claims that these characteristics all create more favourable conditions for religious extremism than standard network organizational patterns. He also argues that religious fanaticism in most cases simply serves as a cover for deeper and more fundamental clan-based hatred. The all-out, total nature of the transnational violent Islamist movement is explained primarily by violent tribalism, rather than by religious extremism per se.

However, it could also be argued that certain elements of network tribalism are more easily traced in the organizational forms of some localized militant–terrorist groups than at the level of transnational superterrorist networks. Indeed, in ethnic groups that are still under

---

[244] On the clan as an organizational form see e.g. Ouchi, W. G., 'Markets, bureaucracies and clans', *Administrative Science Quarterly*, vol. 25, no. 1 (Mar. 1980), pp. 129–41.

the influence of clan traditions (such as Chechens in the North Caucasus), clan affinity is often intertwined with ethnic affinity. Together, they may prove to be more effective than religion as an instrument for mobilizing violence, especially at the early stages of the conflict.[245] Nonetheless, in the organizational patterns, tactics and cultures of the armed ethno-separatist movements active in those regions, elements or remnants of pure traditionalism are much less evident than manifestations of distorted or traumatic modernization. In societies that are dominated by tribal structures (for instance, in the 'tribal belt' across the Afghanistan–Pakistan border), tribalism may directly merge with religious affinity and religious extremism, as in the case of Deobandi Pashtun tribal militias.

In sum, while elements of network tribalism are more likely to be found at the localized level, they are not sufficient to explain the organizational patterns of violence even at this level. It would be an even greater simplification, if not a mistake, to reduce the transnational violent Islamist movement active at the global level to network tribalism. The post-al-Qaeda movement cannot be simply interpreted as an essentially archaic, traditionalist structure based on family–kin clan relationship that skilfully and selectively exploits possibilities offered by postmodern network organizational forms.

First, the concept of network tribalism does not pay full credit to the important role of a common ideology as the main integrating force that ties various cells into a transnational network, even in the absence of formal organizational links. According to the network tribalism concept, clan provides a more solid basis for network links and relationships. Advocates of the concept have even argued that, for instance, the vision of 'jihad' propagated by al-Qaeda and its followers is more in line with aggressive tribalism than with Islamic extremism. This view underestimates the role of Islamist quasi-religious ideology as a driving force for the post-al-Qaeda movement and degrades the ideological imperative to secondary importance. Modern or, to be more precise, postmodern networks do not just imply a high degree of ideological integration, they require it. In contrast, the members of a clan structure do not even have to be ideologically like-minded. Clans are based on ties of a different nature. Another specific feature of all—both violent and non-violent—radical Islamic groups

---

[245] This is true even though religious extremism can rapidly gain force in the course of the armed confrontation.

and movements is the extent to which the Islamist ideology affects all aspects of their activities, including their organizational forms. In other words, the structures of such groups are in many ways a progression and projection of their ideology.

Second, an argument often invoked in defence of network tribalism being the organizational basis for the transnational violent Islamist movement is the fact that some of al-Qaeda's leaders found refuge in areas dominated by clan and tribal relations. The areas most commonly mentioned are the Taliban-controlled parts of Afghanistan and areas along Pakistan's border with Afghanistan (including the Federally Administered Tribal Areas and parts of the North-West Frontier Province). A counter-argument can be easily made that Osama bin Laden and some of his close associates were not necessarily based in Afghanistan and, prior to that, in Sudan because of the spread of clan forms of social organization there. Instead, the leaders of al-Qaeda found refuge in Sudan and Afghanistan primarily because radical Islamist regimes were in power in both countries at the time. Furthermore, in contrast to the times of anti-Soviet 'jihad' in Afghanistan, the modern transnational post-al-Qaeda movement appears to find it easier to recruit volunteers in Muslim diasporas in the West than in remote tribal areas.[246]

Third, an excessive focus on network tribalism may be an attempt to artificially archaize the post-al-Qaeda movement. It ignores the fact that, unlike classic clan structures, modern transnational terrorist networks are not tied to a specific, strictly defined territory. Bin Laden and his closest associates—such as the late Abu Musab al-Zarqawi or Ayman al-Zawahiri—do not resemble clan sheikhs. Nor are they military commanders or political leaders in the traditional sense. Above all, they are typical, almost archetypal, network inspirers. 'Segmental warfare' as described by Ronfeldt[247]—that is, a tactic of loosely coordinated attacks by multiple cells, or segments—is not an exclusive prerogative of traditionalist clans either. It is also effectively waged by modern functional–ideological networks such as certain radical environmentalist movements.

---

[246] In Western Muslim diasporas of various ethnic and national backgrounds the people closely integrated into either relatively archaic, non-modernized clan structures or the mainstream, established religious communities rarely make active members of the post-al-Qaeda movement's cells. See section IV below.

[247] Ronfeldt (note 243).

In the end, an impression may be left that attempts to reduce the post-al-Qaeda movement to network tribalism are at least to some extent dictated by political imperatives. The network tribalism concept apparently builds on experience of Western interventions in the post-11 September 2001 world and, in particular, involvement in Afghanistan since 2001 and Iraq since 2003. After September 2001 there was a disproportionate rise in anti-Islamic rhetoric in the USA and some other Western states.[248] Deteriorating relations with the Muslim world were further aggravated by the war in Iraq. The shift of focus in the studies of organizational forms of superterrorism from ideologically driven networks to network tribalism may have reflected a trend within the US expert community towards a certain strategic adjustment. Part of the US politico-security establishment became increasingly concerned about the negative implications and misleading nature of policy explicitly or implicitly directed towards confrontation with significant parts of the Muslim world.[249] This has stimulated a desire in these circles to temper anti-Islamic rhetoric, 'replace' the threat of Islamic extremism with the threat of clan atavism and attribute the growing level of global terrorist activity primarily to barbaric, archaic and aggressive tribalism. This spirit permeates most of the practical, policy-relevant recommendations made by network tribalism theorists to the US Government and to the USA's allies. One suggestion is, for instance, to draw a strict distinction between the strategy to fight radical Islam and the strategy to confront tribal and clan extremism.[250]

The transnational violent Islamist movement with a global agenda cannot, however, be artificially degraded to the level of tribal clashes in Afghanistan or inter-communal tensions in Iraq. The post-al-Qaeda movement is a far more modernized phenomenon. The most active violent Islamists that form semi-autonomous or self-generating cells with a transnational agenda not limited to any localized context are

---

[248] For an overview see e.g. Human Rights Watch, 'United States—"We are not the enemy": hate crimes against Arabs, Muslims, and those perceived to be Arab or Muslim after September 11', vol. 14, no. 6 (G) (Nov. 2002), <http://www.hrw.org/reports/2002/usahate/>. For a typical example of post-11 September 2001 anti-Islamic rhetoric and a critique of it, see, respectively, Emerson, S., *American Jihad: The Terrorists Living Among Us* (Free Press: New York, 2002); and Muslim Public Affairs Council, 'Counterproductive counterterrorism: how anti-Islamic rhetoric is impeding America's homeland security', Dec. 2004, <http://www.mpac.org/article.php?id=354>.

[249] Of course, 'the Muslim world' cannot be seen as a single entity.

[250] See 'Preliminary implications for policy and strategy' in Ronfeldt (note 243).

not tribal leaders—most of them are educated Muslims with a middle class background. In sum, the main terrorist threat to the West originates not so much in the heart of backward, unmodernized societies with the remnants of tribal and clan structures. Rather, it comes from the most rapidly modernizing areas with the closest contact with the West and from radical segments of Muslim diasporas in the Western states themselves.

IV. Strategic guidelines at the macro level and social bonds at the micro level

Clearly, network characteristics alone are important but insufficient for an organization to be able to wage an effective asymmetrical confrontation at the global level. Attempts to revise network theory as applied to the transnational violent Islamist movement by reducing it to network tribalism are not satisfactory either.

The post-al-Qaeda transnational violent Islamist movement is not a pure network. Like most structures, it also displays certain elements of hierarchy. For instance, it has some leaders—even if they are not necessarily leaders in the classic sense of the word. It both displays informal horizontal links between some of its multiple cells and is a multi-level system that requires at least some vertical links to connect its different levels. This hybrid form allows network and hierarchical elements to reinforce their comparative strengths and compensate for their mutual weaknesses. However, even this hybrid form cannot explain why autonomous cells manage to act in an effective and seemingly coordinated manner in line with the general strategic guidelines formulated by the movement's leaders and ideologues.

An alternative explanation—the concept of leaderless resistance—does not accurately describe the transnational violent Islamist movement either. This concept was developed in the 1980s and 1990s by the US right-wing white-power extremist Louis Beam.[251] Leaderless resistance, which is employed by many right-wing extremists and radical environmentalists, is by definition quite unstable and is not necessarily an effective organizational principle. Leaderless resistance

[251] Beam, L., 'Leaderless resistance', *The Seditionist*, no. 12 (Feb. 1992), <http://www.louisbeam.com/leaderless.htm>, pp. 1–7. For an analysis of the concept see Kaplan, J., 'Leaderless resistance', *Terrorism and Political Violence*, vol. 9, no. 3 (autumn 1997), pp. 80–95.

often serves as a tool of last resort to sustain terrorist activity in the absence of any public support for a radical political programme. It may easily degrade to sporadic, semi-anarchist violence.[252] How can unity of action and strict implementation of generally formulated goals be ensured in a fragmented, dispersed structure? How can they be provided in the absence of a centralized system of direct control and subordination and in a way that prevents them from slipping into meaningless, sporadic and diffuse violence?

Ideological–strategic guidelines at the macro level

The questions put above are not easy to answer. The task is complicated by constraints imposed on analysts working within the organizational network or social network theoretical frameworks by their respective theoretical approaches. As is often the case, a synthesis of the two approaches appears to be more productive and promising from an analytical point of view, particularly as they have both generated valuable insights into the issue of concern.

This need to mix the two approaches is reinforced by the fact that some of the characteristics of modern hybrid terrorist networks, especially transnational networks, are not typical of either pure network or pure hierarchical organizational forms. One of the main specific features of the post-al-Qaeda multi-level network is its ability to ensure effective coordination of actions undertaken by lower-level semi- or fully autonomous cells. This coordination is carried out neither by means of centralized control (as in hierarchies) nor through mutual agreements, compromises and consultations (as in networks).[253] Rather, the movement's activities are coordinated directly by means of strategic guidelines formulated by its leaders and ideologues in a very general way. Among other characteristics that do not exactly fit into either network or hierarchical form is the informal nature of both horizontal links between various units operating at the same level and the vertical links between the different levels. Remarkably, despite the

---

[252] See e.g. Garfinkel, S. L., 'Leaderless resistance today', *First Monday*, vol. 8, no. 3 (Mar. 2003), <http://firstmonday.org/issues/issue8_3/>. This type of violence was known as 'motiveless' terrorism in late 19th century Russia.

[253] On the non-network and non-hierarchical characteristics of modern terrorism see Mayntz, R., *Organizational Forms of Terrorism: Hierarchy, Network, or a Type Sui Generis?*, Max Planck Institute for the Study of Societies (MPIfG) Discussion Paper no. 04/4 (MPIfG: Cologne, 2004), <http://edoc.mpg.de/230590>.

ulterior nature of these links, it appears that they can be effectively and promptly operationalized as required—for instance, to carry out a terrorist attack that involves more than one cell.

Clearly, such effective coordination is only possible if the movement's units, cells, leaders, and rank and file not only support its ideological goals but fully identify themselves with these goals. However, even in that case, within a system where cells are informally interlinked or completely autonomous, strategic coordination through generally formulated guidelines can only be effective provided that the ideology that ties the system together meets certain conditions.

A unity of ideology and strategy may only be achieved if the ideology itself serves as a set of direct strategic guidelines and already contains specific tactical instructions or recommendations. For this to occur, at least two requirements must be met. First, the movement's ideological goals should be formulated in such a way that they may be implemented through various means, in different contexts and circumstances. Whenever an opportunity to undertake violent activities in the name of these goals presents itself, these actions would still qualify as being directed towards the achievement of the ultimate goals. Second, despite the multiplicity of leaders, varying ideological guidance and diversity of organizational forms, a consolidated ideological–strategic discourse needs to be developed within the movement. The overall level of consolidation of the movement's ideology and strategy should thus be unusually high.

Violent Islamist extremism in its most ambitious, globalized form and with its main ideological pillar—the concept of 'global jihad'—is unique in that it manages to meet all of the above requirements. The ideology of radical Islamism that encourages the use of violence through 'jihad' for the sake of its ultimate goals already contains detailed recommendations for practical action. An example is provided by Qutb's recommendations on the formation and activities of the vanguard Islamic revolutionary groups that are—knowingly or unknowingly—followed by the emerging cells of the modern transnational Islamist movement.[254] The more crude popularizers of this ideology, such as bin Laden, have gone further in emphasizing the ideology's encouragement, advance approval of and blessing for any context-specific violent actions, including terrorist attacks. An illustration is provided by bin Laden's 1998 fatwa that prescribed a course

[254] See chapter 3 in this volume.

of action that, regardless of the exact context, circumstance or pretext, would qualify as being directed towards the same 'general goal'. In this fatwa, bin Laden stressed the need 'to kill Americans and their allies' as 'an individual duty for every Muslim who can do it in any country in which it is possible to do it'.[255]

The second requirement for the consolidation of ideology and strategy to the point where they can serve as an effective coordination mechanism for a loosely structured movement is the standardization and unification of strategic discourse. For the post-al-Qaeda movement, with its multiple leaders, ideologues and hybrid, diverse and multi-level organizational patterns, the key role in meeting this requirement has been played by information and propaganda activities. These activities build on—and further develop—the original al-Qaeda ideology. They are increasingly—almost overwhelmingly—conducted through electronic information and communications systems, especially the Internet. Since the mid-2000s, the information providers associated with the transnational violent Islamist movement have qualitatively upgraded and intensified their activities in what appears as an increasingly coordinated way.[256] Intensive online discussions and propaganda have become the main means for ideological–strategic unification for the radical Islamist 'Internet scholars', who range from the first-generation Afghan veteran Abu Yahya al-Libi to many younger clerics, such as the Kuwaiti Hamed bin Abdallah al-Ali. In an attempt to speak as a voice of the collective discourse of 'global jihad' and to reinforce the movement's doctrinal unity, al-Ali, for instance, published the Covenant of the Supreme Council of Jihad Groups in January 2007.[257]

However, even these ideological and doctrinal characteristics of the transnational violent Islamist ideology and strategic discourse cannot dispel the remaining doubts. The question is whether this quasi-religious ideology—even in unity with strategy—could suffice to ensure effective coordination of the movement's activities at the micro level of individual semi- or fully autonomous cells.

---

[255] Bin Laden (note 144).

[256] Examples of these providers include the Al-Fajr Media Center, the Al-Sahab Foundation for Islamic Media Publication (an al-Qaeda-affiliated media house), the Global Islamic Media Front and a number of personal websites of the leading radical Islamist clerics and ideologues of the movement and the affiliated Internet blogs and forums.

[257] al-Ali, H. bin A., [Covenant of the Supreme Council of Jihad Groups], 13 Jan. 2007, <http://www.h-alali.net/m_open.php?id=991da3ae-f492-1029-a701-0010dc91cf69> (in Arabic).

Radicalization and group cohesion at the micro level

In order to systematically engage in an asymmetrical armed struggle on a long-term basis in pursuit of common goals and regardless of a particular area of operation, a very high level of mutual—and highly personalized—social obligation is required. This is something that neither network nor hybrid hierarchized network structure can be expected to provide. Generally, the more network elements that a hybrid movement displays, the greater is the impact of social—individual and group—dynamics on the effectiveness of the activities of its individual cells and on their ability to function as part of a broader network. In order to function effectively, a network requires a higher level of interpersonal trust at the micro level of its units than a hierarchy. At the same time, attempts to consolidate and reinforce a network by formalizing its links and streamlining its structure may not only be futile but may also weaken its main comparative advantages.

It is unlikely that the leaders of al-Qaeda and the post-al-Qaeda movement deliberately masterminded an organizational model that would allow them to make up for the structural weaknesses of the network model without undermining its main strengths. Instead, it was an organic process of organizational evolution and adjustment. As a result, a dynamic system evolved. It both displays a high degree of ideological indoctrination and is characterized by much stronger intra-cell social cohesion, interpersonal trust, commitment and obligations at the micro level than any standard impersonalized functional–ideological network. In order to explore the nature of these mutual obligations at the micro level of individual cells, the sociological paradigm and social network theory have to be considered.

Addressing the problem from this angle has its advantages and drawbacks. The main advantage is the specific attention that this approach pays to the sociological and psycho-sociological aspects of the process of gradual radicalization of Muslims into potential members of radical Islamist cells. It also focuses on further radicalization of the cells themselves through intra-group social dynamics. Indeed, the cells of the post-al-Qaeda movement are united not only by ideological proximity, the feeling of being a part of the same network of semi- or fully autonomous units or informal network-type ties. The best available psycho-sociological accounts of modern transnational violent Islamist networks show that some cells and especially members of the same cell are usually also linked by closer personal and intra-group relations.[258] These close social and personal ties are often established before a group or cell joins the transnational movement. These links are not primarily of the clan or family type; they are more often ties based on friendship, shared regional or national background, or common professional, educational and other experience. This common experience may be acquired not only, or even not primarily, in established places of religious worship—such as mosques and religious schools—but in secular universities, engineering and technical schools, through social activities and so on. According to analysis by Marc Sageman, who was the first to put together the available information on the psycho-sociological characteristics and personal background of 150 active 'jihadists', friendship played an important role for 68 per cent of them. Kinship and family links played the same role for about 14 per cent.[259]

As noted above, the most favourable environment for breeding potential volunteers to join or form cells of the transnational violent Islamist movement appears to form where there is the closest and most intensive contact with 'aliens'. This occurs both in the areas of extended Western economic, military, political and cultural presence and influence in the Muslim world and in parts of Muslim diasporas and communities in the West. With the rise of the Islamist terrorist threat of the post-al-Qaeda type in the West, particularly in Europe, growing attention has been paid by Western analysts to how violent Islamist cells with a transnational agenda emerge. Of particular interest is what factors radicalize Muslims to join this movement. Much of this analysis is dominated by a sociological and psycho-sociological perspective. It appears to be an attempt to rationalize the problem by emphasizing socialization, social integration and intra-group social dynamics as the main factors in terrorism radicalization and recruitment.

---

[258] See e.g. Sageman, M., *Understanding Terror Networks* (University of Pennsylvania Press: Philadelphia, Pa., 2004).

[259] Sageman (note 258), pp. 111–12. Friendship or family ties were an important factor in joining armed jihad for 75% of all the individuals reviewed by Sageman (p. 113). These findings were supported and developed by a broader and more detailed study of the personal background of almost 500 'jihadists'. See Sageman, M., 'Understanding terror networks', Foreign Policy Research Institute E-Notes, 1 Nov. 2004, <http://www.fpri.org/enotes/past enotes.html>; and Sageman, M., *Leaderless Jihad: Terror Networks in the Twenty-First Century* (University of Pennsylvania Press: Philadelphia, Pa., 2007).

There is no need to reproduce in detail the specific mechanisms of cell formation of the transnational Islamist movement. They are context specific, do not conform to a single pattern and have been addressed in other studies (most of which, however, either replicate Sageman's analysis or do not go beyond it in terms of precision and originality).[260] In the most general terms, a group of Muslims of any ethnic and national background get together, establish close friendly relations and form a tightly integrated group. They may range from childhood friends and people originating from the same area in their home countries to Western-born people of the same neighbourhood, university friends or colleagues. This relatively narrow brotherhood of like-minded friends and comrades linked by closely personalized social network ties gradually becomes increasingly politicized. It becomes radicalized under a combination of external—political, psychological or socio-cultural—pressures and internal group dynamics, and finds natural guidance and ready answers to many of its concerns in the radical Islamist ideology. At some point, group members realize the futility of mere talk and the need to turn to active propaganda by deed. The group is then ready to become an integral, semi- or fully autonomous part of the transnational Islamist movement, often joining it as a cell.

As for the more specific radicalization and cell-formation mechanisms, they may be significantly nuanced even for different kinds of diaspora Muslims in the West. Islamist cell members range from visitors and first-generation immigrants to second- and third-generation European-born citizens or even, in some cases, Western converts. The same applies to how they finally join the transnational movement. For some cells the direct link to 'jihad' through a contact with an active, preferably veteran 'jihadist' is necessary. Some analysts, especially in early post-11 September 2001 years, even saw 'the accessibility of the link to jihad' as the most critical element in the entire chain.[261] There is not one single pattern, however. Some cells now appear to see direct action itself as the quickest and most access-

[260] See e.g. Taarnby, M., 'Understanding recruitment of Islamist terrorists in Europe', ed. M. Ranstorp, *Mapping Terrorism Research: State of the Art, Gaps and Future Direction* (Routledge: London, 2007), pp. 164–86; and Bokhari, L. et al., *Paths to Global Jihad: Radicalisation and Recruitment to Terror Networks*, Norwegian Defence Research Establishment (FFI) Report no. 2006/00935 (Norwegian Defence Research Establishment: Kjeller, 2006), pp. 7–21.

[261] Sageman (note 258), pp. 120–21.

ible way to become part of the broader movement, find ways and means to organize terrorist activity on their own, and carry out terrorist acts. In other words, while a direct link to existing cells or the leaders of the transnational violent Islamist movement may be an important condition for an individual cell to engage in terrorist activity, it is not critical. The duration of the radicalization and cell formation process may also vary. Radicalization patterns also change over time. Earlier analyses described it as a long and gradual process requiring personal intercommunication.[262] More recent sources point to an increasingly rapid radicalization of Islamist terrorist cells in, for example, Europe.[263] Much of this increasingly rapid radicalization is enabled and facilitated by the growing role of online communication through electronic information providers and Internet blogs and forums.[264]

At the same time, the excessive focus on sociological aspects of Islamist radicalization in the West and on social alienation and group dynamics as the main explanation of the formation of Islamist terrorist cells has its drawbacks. Intentionally or unintentionally, it tends to depoliticize terrorism—perhaps, the most politicized of all forms of violence. It downgrades the importance of broader international political agendas and their quasi-religious interpretations for violent Islamists in Europe and around the world. By prioritizing the mechanisms of radicalization, this approach often overlooks or de-emphasizes the more important motivations and driving factors behind the formation of post-al-Qaeda Islamist terrorist cells. These factors may have little to do with problems of socialization, lack of social integration, immediate social circumstances and social group dynamics. This is particularly true for those Islamist terrorists who, unlike some of the poorly integrated recent immigrants, may be well-integrated second-generation citizens of European countries or even European converts to Islam.[265] In contrast, the radicalization process of visitors and first-

---

[262] See e.g. Sageman (note 258), p. 108; and Taarnby (note 260), p. 181.

[263] See e.g. Europol (note 11), pp. 1, 18–19.

[264] See e.g. Sageman, M., 'Radicalization of global Islamist terrorists', Testimony before the US Senate Committee on Homeland Security and Governmental Affairs, 27 June 2007, <http://hsgac.senate.gov/index.cfm?Fuseaction=Hearings.Detail&HearingID=460>, p. 4.

[265] E.g. no second-generation citizen could be better integrated than Mohammed Sidique Khan, the leader of the Leeds group responsible for the London bombings of 7 July 2005. According to the available official data, the members of this group had 'largely unexceptional' backgrounds, with little to distinguish their formative experiences from those of many others of the same generation, ethnic origin and social background. All this points to the

generation migrants may involve a considerable degree of social isolation and identity crisis resulting directly from a sharp socio-cultural shift, such as immigration.

In sum, there is no single or simple social or radicalization pattern for members and cells of the transnational violent Islamist movement in the West and elsewhere. The nature of their quasi-religious, politico-ideological radicalization and cell organization is not always and not necessarily a product of their own poor social integration. For example, their negative formative socio-cultural experiences in the West may be reinforced by social group network dynamics. This combination may play a role in preparing them to turn to violent actions against civilians in what they believe is the cause of fellow-Muslims suffering around the world. However, they primarily frame their actions in quasi-religious, political, almost neo-anti-imperialist discourse driven by what they see happening in Afghanistan, Iraq and elsewhere. It is always a combination of a feeling of alienation from the 'imperfect', 'immoral', 'corrupt' (*jahiliyyah*) society that surrounds them with the strong mobilizing impact of international political and military developments, such as those in Afghanistan and Iraq. The impact of these events, seen as injustices and crimes against 'all fellow-Muslims', is reinforced by and reinterpreted through the prism of radical Islamist ideology.[266]

As is always the case with violent Islamism in general, and cells of the post-al-Qaeda transnational movement in particular, an analysis of their organizational patterns must return to their quasi-religious ideology. As has been noted by other observers, the movement appears to

'potential diversity of those who can become radicalised'. British House of Commons, *Report of the Official Account of the Bombings in London on 7th July 2005* (The Stationery Office: London, May 2006), <http://www.official-documents.gov.uk/document/hc0506/hc10/1087/1087.asp>, pp. 13–18 etc.; and British Intelligence and Security Committee, *Report into the London Terrorist Attacks on 7 July 2005*, Cm 6785 (The Stationery Office: London, May 2006), <http://www.official-documents.gov.uk/document/cm67/6785/6785.asp>, p. 43.

European converts to Islam are a small minority among terrorists of this type but their existence shows that not all Islamist terrorists in the West are migrants or their descendants.

[266] As noted in Khan's 'suicide bomber farewell videotape', 'until we feel security, you will be our targets' and 'until you stop the bombing, gassing, imprisonment, and torture of my people we will not stop this fight'. The video was broadcast on the Al-Jazeera television channel on 1 Sep. 2005 and is quoted in British House of Commons (note 265), p. 19. In the words of another of the Leeds group of suicide bombers, Shehzad Tanweer, attacks 'will intensify and continue, until you [the USA and allies] pull all your troops out of Afghanistan and Iraq'. Middle East Media Research Institute (MEMRI), 'Al-Qaeda film on the first anniversary of the London bombings', Clip transcript no. 1186, 8 July 2006, <http://www.memritv.org/clip_transcript/en/1186.htm>.

grow and its cells reproduce themselves less through recruitment than though volunteering and the self-generation of cells. What role then do the movement's main leaders and ideologues play in this process? Compared to the anti-Soviet 'jihad' in Afghanistan of the 1980s, this is a largely innovative and effective pattern that involves the spread and use of the extremist ideology as an organizing principle. The movement's upper-level leaders concentrate their energy and resources on spreading and popularizing the movement's ideology. This quasi-religious ideology already contains a direct recipe for violent action among potential sympathizes, some of which—a small minority—later become volunteers to join the movement's cells. A series of dramatic and politically divisive international developments, such as the armed conflicts in Afghanistan and Iraq, provide a favourable context for the spread of violent Islamism and up-to-date illustrations of the main theses of this extremist ideology. These political developments create an atmosphere in which the message of the violent extremists appeals to enough Muslims to provide more than enough volunteers to fight.

V. Conclusions

As shown in this Research Report, an analysis of any aspect of the organizational development of terrorist cells of the transnational Islamist network either involves or ends up with the movement's universalist quasi-religious extremist ideology. This interdependence of ideological and structural aspects is striking and further underscores their inseparability in the analysis of the post-al-Qaeda transnational violent Islamist movement.

This movement's ideology does not in principle favour strictly hierarchichal organizational forms, which it perceives as instruments of the 'enslavement of men by other men' and a manifestation of *jahiliyyah*.[267] The movement retains a strong egalitarian element and gives

---

[267] The same, of course, applies to 'markets' as another classic organizational form. It may be noted that while, for instance, the followers of various left-wing revolutionary ideologies in theory also strongly opposed all forms of exploitation and subordination, that did not prevent them from establishing some of the most highly centralized and the strictest hierarchical systems in the world, at both the state and non-state levels. Against this background, violent anti-system Islamists, at least at the transnational level, appear to have been more consistent in matching their stated beliefs and values with the prevailing organizational forms of their movement.

a general preference to networks over hierarchies. However, it goes far beyond the standard modern ideologically integrated network of the functional type (such as the anti-globalist campaign). It can be more accurately described as a mixed, or hybrid, multi-level structure. Displaying many key network characteristics, as well as some hierarchical elements, it also has several specific features that are not typical of the main known organizational forms. The high degree of informal coordination of this multi-level network's activities outmatches the coordination mechanisms of many far more formalized structures. What makes this possible is a combination of the extremist ideology of violent Islamism at the macro level and the unusually tight group cohesion provided by strong social bonds and obligations at the micro level of individual cells. The latter are not so much traditionalist clan-based entities as they are close associations (or brotherhoods) of like-minded friends and comrades.

Perhaps most importantly, the movement's extremist quasi-religious ideology and increasingly consolidated strategic discourse serve not only as its structural glue but as an organizing principle. It allows individual cells to engage in whatever violent activity they can master at the micro level—regardless of their area of operation—in a way that still makes the perpetrators and the global audience see these activities as coordinated at the macro level and ultimately directed towards the same goal.

6. Conclusions

In the early 21st century, violent Islamism has become the main ideological basis for terrorist activity at the transnational level. It is also one of the main extremist ideologies of groups that use terrorist means in a number of more localized, national contexts. In various academic, political and security quarters much has been said and written about the need to counter Islamic extremism by ideological methods, particularly by using Islam in its moderate version against Islamic extremism. The author of this Research Report has herself faithfully contributed to these well-intentioned but, it now appears, largely futile efforts.

Most proposals of this kind boil down to a set of standard recommendations. For example, they include calls to encourage mainstream Islamic groups, madrasas, charities and foundations both in their practical social, humanitarian and reconstruction activities and in their political, ideological and religious debates with Islamic radicals.[268] These debates are centred on such issues of critical relevance and importance to anti-terrorism as the concepts of martyrdom and jihad. For instance, they encourage the efforts by the moderate Muslim clergy to promote the traditional religious bans on targeting the enemy's women and children (as long as they do not take up arms) and on destroying buildings that are not directly related to a battle.

So far the efforts to use moderate Islam against Islamist terrorism have generally failed to moderate the extremist ideology of violent Islamists. Nor have they helped to curb terrorist activity worldwide, which continues at a dangerous level. Part of the problem is that these well-intentioned efforts are based on an understanding of Islamist terrorist threats at levels from the local to the global that emphasizes their religious nature, rather than their quasi-religious nature. This approach thus overestimates, for instance, the power of theological arguments and the role of moderate clergy in confronting the violent radicals.

More generally, in contrast to radical movements at the more localized level that combine elements of nationalism and Islamism and display varying degrees of pragmatism in their social and political

---

[268] Stepanova, *Anti-terrorism and Peace-building During and After Conflict* (note 20), pp. 45–48.

behaviour, the extremist ideology of the supranational violent post-al-Qaeda movement is in principle unlikely to be moderated. This is a reality that many analysts and practitioners are reluctant to recognize, even though some of the most critically minded may well sense it. Moreover, this universalist Islamist ideology with unlimited goals and transnational outreach will persist. One reason for the ideology's persistence is that it is, in part, a global reflex reaction to, or a symptom of, objective socio-political, socio-economic and socio-cultural processes of the contemporary world—first and foremost, 'traumatic' globalization and uneven modernization. So long as this quasi-religious ideology continues to reflect the radical reflex reaction to these processes, it will continue to spread. In addition, the reflex function of universalist Islamist ideology is reinforced by its role as a more specific reaction to the political realities of the early 21st century. International political developments—such as the conflicts in Afghanistan and Iraq and the continuous impasse in the Israeli–Palestinian conflict—all appear to conform to and reinforce the radical Islamists' alarmist world view.

Against this background, the most that the use of moderate Islam against violent Islamic extremism can achieve—even under the most favourable circumstances and in combination with other socio-economic and political tools—is a limited influence on the radicals' broader support base. It cannot impede or effectively constrain the process of ideological radicalization—at best, it can merely complicate the process.

I. Nationalizing Islamist supranational and supra-state ideology

It is worth recalling that the asymmetry dealt with here is a two-way asymmetry. One party to this asymmetrical confrontation is the state (and the international system in which states, despite the gradual erosion of some of their powers, remain key units). The state is faced with the toughest of its violent non-state anti-system opponents—the supranational, supra-state resurgent Islamist movement of the multi-level, hybrid network type. While the movement's ultimate utopian, universalist goals are unlikely to be realized, it can still spread havoc through its use of radical violent means, such as terrorism and especially mass-casualty terrorism.

Within this asymmetrical framework, the state and the international community of states are incomparably more powerful in the conventional sense—in terms of their aggregate military, political and socioeconomic potential. States also enjoy a much higher formal status within the existing world system and remain its key formative units.

However, the transnational violent Islamist movement inspired by al-Qaeda has its own strengths and comparative advantages in waging an asymmetrical confrontation. This Research Report argues that these asymmetrical advantages of violent anti-system non-state actors employing terrorist means are their extremist ideologies and structures. These comparative advantages are most evident at the transnational, or even globalized, level. This thesis by no means implies that such radical ideologies are generally superior to the mainstream, more moderate ideological frameworks. Nor does it suggest that organizational forms employed by transnational militant–terrorist actors are in any way better than state-based organizational structures dominated by hierarchical forms. It only means that these non-state actors may be better ideologically and organizationally tailored for an asymmetrical confrontation with an otherwise incomparably more powerful opponent.

It follows that, if the international system of states tries to engage in a full-scale conflict of ideologies in the framework of asymmetrical confrontation with violent Islamists (and within this framework alone), then by definition it puts itself at a disadvantage. It is precisely because of the modernized, moderate, relatively passive nature of the mainstream ideologies of state actors that they cannot compete with a radical quasi-religious ideology. They can offer little to compete with Islamist extremism as a mobilizing force in asymmetrical confrontation at the transnational level. In other words, on the ideological front the state and the international system may be faced with a reverse (negative) asymmetry that favours their radical opponents.

It is self-delusional to think that quasi-religious extremism in the form of violent Islamism can be neutralized by using modern Western-style democratic secularism. It cannot even be undermined by the moderate, mainstream currents operating within the same religious and ideological discourse, that is, by moderate Islam. While such efforts do no harm, they simply do not work. They are unlikely themselves to produce the intended result of moderating the ideology of the violent extremists, especially those employing terrorist means.

An ideology, including a highly extremist one, needs to be countered first and foremost at the ideological level: as the Hamas Covenant justly notes, 'a creed could not be fought except by a creed'.[269] This is true, but only as long as the two creeds are at least comparable (or symmetrical) in terms of the power of appeal and mobilization, and perhaps even the degree of radicalism. The radical quasi-religious ideology dealt with here is much more than a marginal religious cult—it has a worldwide spread and appeal. It inspires enough people in different parts of the world to volunteer to join the cells of the transnational Islamist movement and ultimately become part of the larger movement through violence (including self-sacrifice). The weakening, erosion or undermining of such an extremist ideology requires an ideology of comparable strength, coherence and power of persuasion.

In the absence of any equally coherent, mobilizing, universalist and all-embracing moderate ideology to counter supranational violent Islamism on its own terms and at the global level, what conclusions does this reverse ideological asymmetry lead to?[270] The logical conclusion is that the current negative ideological asymmetry that benefits the radical anti-system actors should be adjusted, stimulated or forced to develop in a more symmetrical direction. If that goal cannot be achieved by means of the direct involvement of the state actors that form the international system and their dominant ideologies, then could it be done by others? To put it simply, if a moderate ideology does not work as an effective counterbalance to violent Islamism, then perhaps a more radical one will do better.

The challenge of transnational violent Islamism

To explore potential alternatives, the first question to address is what makes radical Islamism such a powerful ideology in asymmetrical confrontation with 'the system', especially at the transnational level.

First, as repeatedly stressed in this Research Report, Islam in its fundamentalist forms, and especially in its more politicized Islamist versions, is more than a typical religion. Much of this Research Report and other, more specialized works have been devoted to exploring and revealing Islam's dialectic, quasi-religious nature.

---

[269] Covenant of the Islamic Resistance Movement [Hamas] (note 124), Article 34.

[270] On reverse ideological asymmetry see chapter 1 in this volume, section II.

Applying a standard, modern, Western-centred interpretation of religion to Islamist thinking and practice (where religion means 'the system and way of life' in 'all its details'[271]) is questionable. The system and way of life that radical Islamists aspire to build, either by violence or through peaceful means, is not a theocracy. It is not limited to confessional issues and does not aim at forcing people of other confessions to become Muslims against their will. Rather, *imaan* (faith) is viewed as the fundamental mainstay of a holistic concept of a global order that ties together all its inseparable socio-ideological and political aspects and manifestations. *Imaan* is seen as the source of divine laws and rules that provide for a far more just and fair system than 'the rule of men'.

Second, as is clear from the above, supranational Islamism pursues unlimited goals—it is nothing less than a concept of a global system based on God's direct rule. Among other things, this means that its appeal is truly global and its focus is not limited to the West as the main challenge to be overcome on the way to the global Caliphate. It is a widespread delusion among the governments and other institutions of the Euro-Atlantic community, most evident in the USA, that they are themselves the ultimate target of the Islamists. The West is certainly an important opponent of Islamists and, in their discourse, a powerful pathogenic source of *jahiliyyah* of all kinds, but it is not their main or ultimate enemy. The quasi-religious ideological categories with which transnational Islamists operate and their ultimate goals, as well as the more specific reasons for waging violent 'jihad', go far beyond mere confrontation with the West. The need 'To establish the Sovereignty and Authority of God on earth, to establish the true system revealed by God' is seen by radical Islamists as reason 'enough to declare Jihad'.[272] The global and universal nature of these goals is fully realized and acknowledged by Islamist ideologues: 'the subject matter of this religion is "Mankind" and its sphere of activity is the entire universe'.[273]

At this supranational level violent Islamism is at the utmost of its strength. At least since the decay of Marxism and other leftist internationalist ideologies, there has been no other equally coherent protest ideology with an alternative globalist vision (which, for instance, the

[271] Qutb (note 101), p. 231.
[272] Qutb (note 101), p. 240.
[273] Qutb (note 101), p. 242.

modern anti-globalist movement fails to provide). At this level the violent transnational Islamist movement is hardest to counter.

The contrast between this hyper-globalist vision and radical nationalism, no matter how extreme or how narrowly focused on ethnicity, could not be sharper. The violent Islamist movement with micro-cells active in different parts of the world is not simply an internationalist phenomenon or even a transnational or supra-state one. It pursues a truly universalist agenda that does not respect geographical boundaries or bind itself with ethnic, national, state or even confessional limitations. The ideology of this movement can be more accurately described as *non-state*. It does not simply aspire to control most existing states—it rejects and devalues the very notion of the modern nation state, the beginning and end of all types of nationalism. In its most ambitious form, this movement exists, dreams and operates in a dimension that lies outside the state framework. In that dimension, people are characterized and distinguished not by their ethnicity, nationality, origin and so on, but on the basis of whether or not they share the faith in the one God. For radical Islamism, God is the only lord on earth and no nation state, including all existing Muslim states, can substitute for the God-given system of rules and laws. These rules should apply to everyone, regardless of their national, racial or confessional background.

At the global level, such a quasi-religious ideology cannot in principle be reconciled with nationalism of any kind. This is one of the main reasons why modern global superterrorism is dominated by supranational, quasi-religious post-al-Qaeda Islamist networks.

Radical nationalism and religious extremism

Clearly, no other kind of religious extremism contrasts so sharply with radical nationalism and ethno-separatism as the supranational, al-Qaeda-inspired vision does. However, radical nationalism and less transnationalized forms of religious extremism are not necessarily mutually exclusive in a localized context. Islamism may be effectively employed as an additional justification and ideological basis for terrorism as a tactic of armed resistance in national liberation (e.g. as in Iraq and the Palestinian territories) or ethno-political, separatist conflicts (e.g. as in Chechnya, Kashmir and Mindanao). Nonetheless, when employed in these contexts, Islamism as an essentially transnational, quasi-religious ideology of a universalist type has to adjust, transform and nationalize itself. This is the only way in which it can bridge the gap between its supranational vision and the radical nationalists' obsession with the nation state.

Furthermore, under certain circumstances a resort to violent Islamism can reinforce radical nationalist, national liberation or ethnonationalist movements in Muslim-populated areas. An Islamicization of what originally emerged as a broadly nationalist or ethno-separatist movement can serve as an additional source of public mobilization and legitimization for an armed non-state actor. It can also effectively back a nationalist agenda with one of the Islamists' key assets and strongest advantages—their broad networks of social–humanitarian support for the population. Perhaps more importantly, Islamicization allows a localized armed group to extend its audience and its potential support base by appealing, theoretically, to the entire *umma*. In a more practical sense, it can at least reach out to similar-minded Islamist extremist groups and networks around the world. This is in contrast to secular ethno-nationalists, who cannot count on any major public support beyond the ethnic group in whose name they claim to act and whose interests they claim to defend.[274] A minor exception are their links to like-minded armed ethno-separatist groups in the breakaway regions of other countries.

However, from the point of view of using ideological means to weaken, or even neutralize, the ideological basis of terrorist activities, special attention should be paid to the ways in which religious extremism and ethno-nationalism may also weaken one another.

First and foremost, a combination of Islamism with nationalism may narrow the transnational Islamist agenda by tying it to a national context. This would make it more focused on concrete, pragmatic and far more achievable goals in specific regional, national or local political contexts. Such a nationalization of transnational Islamist movements is not an uncommon phenomenon. It normally occurs for pragmatic, rather than purely ideological, reasons. Examples include Hezbollah and the former Gaza branch of the Muslim brotherhood that became Hamas. Hamas has become almost synonymous with Palestinian nationalism and effectively competes with secular Palestinian organizations on that count. The growing nationalization of such

---

[274] See Stepanova, *Anti-terrorism and Peace-building During and After Conflict* (note 20), pp. 46–47.

Islamist groups is a way for them to gain more solid public legitimization, to develop a stake in national politics and to increasingly and progressively rationalize their agenda by entering into the mainstream political process. The greater the role of pragmatism in the strategies and practices of such nationalized, or nationalizing, Islamist movements, the greater the gap between their activities and their ultimate (and unrealistic) quasi-religious, transnational goals. This gap makes them more amenable to the rational influences, pressures and constraints that remain the main instruments at the disposal of states and international organizations.

Second, Islamicization of what was previously a predominantly nationalist movement may often be counterproductive from the perspective of achieving the movement's ethno-nationalist, including separatist, goals. By associating themselves with violent Islamism, especially of an explicitly transnational nature, ethno-nationalists run the risk of eroding the support for their main original goals. These goals, such as autonomy, self-government or independence, were what originally earned them support from the ethnic group in whose name they claimed to act and use violence. Not all of the supporters, sympathizers or those who are indifferent to the movement's original nationalist agenda, but make no effort to oppose it, would be ready to back a broader, transnational agenda of violent Islamism.

This has been demonstrated by the evolution of radical ethno-nationalism in the North Caucasus region of Russia in the late 20th and early 21st centuries. During the 1990s it evolved from a nationalist movement to a movement combining radical nationalism with violent Islamism. On the one hand, the radicalization of Islam and the Islamization of the Chechen resistance served as additional mobilization tools. Islamization also helped the rebels to regionalize the conflict (to reach across ethnic barriers) and allowed them to appeal for financial and political assistance from foreign Islamist organizations. On the other hand, Islamic extremism not only failed to gain a mass popular following in Chechnya but might have reduced the appeal of the ethno-nationalist separatist cause among some Chechens. In particular, some were dismayed by the militants' attempts to impose elements of sharia and did not want to live in a Taliban-style Islamic state. The rise of radical Islam in Chechnya has also weakened the resistance by provoking a series of violent splits within the armed movement between radical Islamists and the more traditional nationalists following moderate Sufi Islam. It may also have been a factor in the decrease in financial support from the Chechen diaspora elsewhere in Russia to the movement as it became increasingly Islamist and transnationalized its agenda.[275]

In order to erode the strengths of Islamist supranationalism at a national level, there are few workable alternatives to using nationalism. The nationalization of transnational violent Islamism can at least make the latter more pragmatic and, thus, easier to deal with. Radical nationalism in its different forms seems to be the only ideology that is radical enough for this purpose, especially in the context of an ongoing or recently ended armed confrontation. This role can be effectively played by both the more narrow ethno-separatist movements and the broader nationalist resistance movements, including cross-ethnic and cross-sectarian ones (such as in the Iraq and Israeli–Arab conflicts).

In sum, nationalism, especially cross-confessional or multi-ethnic nationalism, is no less powerful an ideology in a local or national context as supranational quasi-religious extremism. It can be employed as a way to weaken some of the most dangerous characteristics and erode some of the main comparative advantages of transnational violent Islamism with a global outreach. It can be particularly effective in helping degrade and refocus the terrorists' agenda by regionalizing or localizing it.

From secular civic nationalism to confessional nationalism

The mainstream modern state ideologies in Western and some developing non-Western states are based on liberal, market-oriented democracy, sometimes with a moderate socialist bent. In much of the rest of the world, mainstream ideologies are also represented in the varying forms of national modernism, whether of a more secular or a moderate religious bent. More radical ideologies with a strong capacity to mobilize socio-political protest are more commonly employed by non-state actors, especially in times of conflict. Of these ideologies, radical nationalism, whether confessional or cross-confessional,

---

[275] This is not to mention that the adoption of transnational Islamist rhetoric by Chechen terrorist groups (including slogans borrowed word for word from Osama bin Laden, such as 'we want to die more than you want to live') greatly facilitated Russia's effort to integrate the war in Chechnya into the US-led 'global war on terrorism'.

appears to be powerful enough to effectively match quasi-religious violent Islamist extremism at and below the level of the nation state. Nationalism by default lacks transnational, let alone global, appeal. However, in a localized context, such as local and regional armed conflicts involving Muslim populations, it can outmatch religious extremism—especially in its purely transnational forms—in its mobilization power and capacity. While radical nationalism may also effectively play the role of protest ideology and may oppose the status quo through violent means, it is not an irreconcilable anti-system ideology at the transnational level. Its protest role is by definition limited to a national context. Radical nationalists do not pretend to exist in an entirely different, parallel dimension. Instead, they are determined, while challenging particular states, to not only recognize but even prioritize the state as one of the central elements of the world system and to focus their agenda on the creation, restoration or liberation of a state.

A word of caution is needed here to warn against simplistic interpretations, such as a picture of an ideological 'front', or 'battleground', where the lesser of the two evils must be chosen. Nor is this a call to revive types of nationalism that may not be relevant in the context of many modern armed conflicts. Attempts to artificially construct modern civic nationalism from the outside in areas where it has little domestic foundation (such as Afghanistan) are inadequate. Efforts to revive the outdated models of the left-wing secularized, often anti-colonial, nationalism that was the driving force of many of the violent protest movements of the 20th century would not work either. Today's radical nationalism is a nationalism of a different kind. It is less secularized and is more reliant on confessional elements as an additional, and powerful, way of reinforcing national and cultural identity in an increasingly globalizing and less ideological world. In a multi-ethnic and multi-confessional environment, this role should ideally be played by cross-confessional nationalism.

In sum, the news about the end or retreat of nationalism, both in general and as one of the main ideologies of armed non-state actors in particular, may not even have been premature. It may simply be inaccurate. Nationalism is not gone—it is simply changing form. The present era is characterized by the dynamic interaction of conflicting—and interdependent—trends of globalization and fragmentation, of universalism and the rise of identity politics. In this context, radical nationalism as a protest ideology is not a replication or a distorted mirror image of a Western-style secular nationalism. Nationalism of that type was associated with a certain period of anti-colonial struggle, has in many ways discredited itself in many parts of the Muslim world in particular and has been perceived, especially by Islamists, as an 'alien' phenomenon. The new type of nationalism is a more indigenous and less secular one and is more closely and intimately tied to— and shaped by—specific local, national and regional contexts.

II. Politicization as a tool of structural transformation

As noted above, extremist ideology as a mobilizing, indoctrinating power is not the sole comparative advantage of terrorism, including transnational Islamist terrorism. Other important advantages of non-state actors employing terrorist means usually include their structural models and organizational forms. The two main comparative advantages of non-state actors waging asymmetrical confrontation at either sub-state or transnational level—their ideologies and structures—are interconnected. They should be analysed and addressed in concert.

It should be noted that, in the case of violent Islamist movements, the interdependence between extremist ideology and the organizational forms tailored to advancing its goals is a particularly unequal one. The impact of Islamist ideology on the movement's structure is far greater than that of the structure on the ideology. The more radical is the Islamist ideology of a movement, the higher the degree to which every aspect of its structure and activity, including its organizational forms, is subordinated to its ideology. This interdependence is not a one-sided relationship, however. The organizational system of the transnational violent Islamist movement also develops as an organic multi-level hybrid network. This development is coordinated by general strategic guidelines formulated by the movement's leaders and ideologues but is also reinforced by close personalized brotherhood ties at the level of its micro-cells.

A state that wishes to effectively normalize and streamline the structural capabilities of violent movements that it cannot defeat militarily must also adjust its own organizational forms in response to this challenge. Such an adaptation may help to neutralize some of the comparative structural advantages of non-state actors in asymmetrical confrontation. This task requires a range of strategies and approaches including, for example, the introduction of some elements of network organizational design into relevant state security structures (e.g. by means of more active inter-agency cooperation). However, in making these changes, the state must make sure that it does not lose its own comparative structural advantages.

However, in some states—usually weak or not fully functional—a more politically controversial option is apparently being followed: making a symmetrical response to an asymmetrical terrorist threat. Such a response involves non-state groups that are loosely affiliated with the state or act in its support without its formal approval. While some of their goals may differ from those of the state, they also usually have a strong self-interest in acting in line with the state's interests. These state-affiliated and pro-state actors follow organizational patterns similar to those of the state's main asymmetrical protagonists, thus depriving the latter of one of their key advantages.

A second task that a state should undertake in order to normalize the structure of a violent movement is to try to formalize the informal links within the opponent's organization. This is no less important and is more challenging than the first task of the state adapting itself, and the two should ideally be coordinated. All possible efforts must be made to turn relatively decentralized terrorist networks into more formal, more streamlined and more hierarchized hybrids.[276] This goal can be best achieved by encouraging the politicization and political transformation of the major armed groups that employ terrorist means and the general demilitarization of politics, especially in post-conflict areas. That implies stimulating the armed groups to become increasingly politicized and involved in non-militant activities. They should be encouraged to form distinctive and fully fledged political wings (rather then merely civilian 'front organizations' for fund-raising and propaganda purposes). These political wings could then gradually develop a stake in increasing their legitimization, and so develop into or join political parties and eventually be incorporated into the political process. The evolution of Hezbollah provides an example of the transformation of a radical armed Shia group. Having been created for the purposes of armed resistance to the Israeli occupation of southern

---

[276] At the national level and in the context of more localized asymmetrical armed conflicts, this imperative becomes all the more pressing at the stage of peace negotiations, as the structural model typical for many of these groups complicates centralized strategic decision making and coordination of actions by their different elements, calling into question their adherence to any formal or informal agreements that could be achieved.

Lebanon by all possible means, including terrorism, it gradually became increasingly involved in social work and political activities and became the main political representative of Lebanon's Shia Muslims, the country's largest—and growing—community.[277]

Such an evolution of a violent non-state group into a legal political party could be extremely difficult and may be preceded by or lead to violent splits and the intensification of internal and sectarian violence. It may even drive more radical factions to more actively resort to terrorist means, in an increasingly irrational manner. Despite this, it is the best way to widen the gap between the more moderate elements within an armed opposition movement, who can be demilitarized and included in the political process, and the more radical underground hardliners. It allows the most extreme hardliners to be more easily isolated, marginalized, delegitimized and, ultimately, forced to stop fighting or relocate to other countries (as was the case for many off-shoot groups of the PLO and the PFLP). The dissolution or destruction of the hardliners could then be more easily achieved through a combination of more specialized counterterrorist techniques and military means. In sum, while the process of political transformation would not necessarily result in a group's rejection of violence once and for all, it could stimulate it to abandon the most extreme violent tactics such as terrorism and facilitate and contribute to the marginalization of its most radical elements.

III. Closing remarks

There appear to be very few effective ways to deprive the trans-national violent Islamist movement of its most dangerous and far-reaching ideological advantages (such as its globalist, all-embracing, supranational nature, goals and agenda). One of these ways is a politically costly, relatively unorthodox and extremely time- and energy-consuming strategy aimed at openly or tacitly encouraging the nationalization of its ideology. At the very least, this transformation process should not be hampered when it occurs in a natural and organic manner.

This approach should be supplemented with and paralleled by efforts aimed at the politicization and political transformation of vio-

[277] See also notes 208 and 209.

lent Islamist movements in a specific national context. These efforts should be seen as a way of making violent Islamists normalize their organizational and structural forms around a more concrete, specific and nationalized political programme. This in turn should help formalize informal or semi-formal links within the organization and create an identifiable leadership and political bodies that could be focused on and dealt with. This seems to be the most direct and realistic way of dealing with an elusive multi-level network of semi- or fully autonomous cells, effectively coordinated through general, quasi-religious guidelines by dispersed leaders and ideologues. The goal should be to turn it into a more normal organizational system that loses some of it key network and other structural advantages in an asymmetrical confrontation.

Unless transnational violent Islamism is first nationalized and then transformed in both ideological and organizational terms through its co-optation into the mainstream political process, it is highly unlikely to become amenable to persuasion. It is, indeed, unlikely to be susceptible to any external influence. It is even less likely to be crushed by repression, which it actually thrives on. In this sense, the most radical and the most perilous supranational al-Qaeda-inspired breed of violent Islamism is practically invincible, as its converts do not defend a territory, nation or state. They fight for an all-embracing mode of existence, a way of life, a holistic and global system through the establishment of the 'direct rule of God on earth' which, as they genuinely believe, would guarantee the freedom of human beings from any other form of governance.

Select bibliography

I. Sources

Data collections

University of Maryland, Center for International Development and Conflict Management (CIDCM) data resources on peace and conflict:

CIDCM Minorities at Risk Project database, <http://www.cidcm.umd.edu/mar/data.asp> .

Hewitt, J. J., Wilkenfeld, J. and Gurr, T. R., CIDCM, *Peace and Conflict 2008* (CIDCM: College Park, Md., 2008); the executive summary is available at <http://www.cidcm.umd.edu/pc/>.

Marshall, M. G. and Gurr, T. R., CIDCM, *Peace and Conflict 2005: A Global Survey of Armed Conflicts, Self-Determination Movements, and Democracy* (CIDCM: College Park, Md., 2005), <http://www.cidcm.umd.edu/publications/publication.asp?pubType=paper&id=15>.

Federation of American Scientists (FAS), 'Liberation movements, terrorist organizations, substance cartels, and other para-state entities', <http://www.fas.org/irp/world/para/>.

Iraq Body Count, <http://www.iraqbodycount.org/>.

Jane's Terrorism and Insurgency Centre, <http://jtic.janes.com/>.

Memorial Institute for the Prevention of Terrorism (MIPT), Terrorism Knowledge Base, <http://www.tkb.org/>, including data from the RAND Terrorism Chronology and the RAND–MIPT Terrorism Incident Database.

Monterey Institute of International Studies, Center for Non-Proliferation Studies (CNS), WMD Terrorism Chronology and Conventional Terrorism Chronology, <http://cns.miis.edu/research/terror.htm>.

United Nations Security Council, 1267 Committee, Consolidated list of individuals and entities belonging to or associated with the Taliban and al-Qaida organisation, <http://www.un.org/sc/committees/1267/consolist.shtml>.

University of Bergen, Terrorism in Western Europe: Event Data (TWEED) data set; see Engene, J. O., 'Five decades of terrorism in Europe: the TWEED dataset', *Journal of Peace Research*, vol. 44, no. 1 (Jan. 2007), pp. 109–21.

Uppsala University, Uppsala Conflict Data Project, Global Conflict Database, <http://www.pcr.uu.se/database/>.

US Department of State, Office of the Coordinator for Counterterrorism, *Patterns of Global Terrorism* (annual, until 2003) and *Country Reports on Terrorism* (annual, since 2004), <http://www.state.gov/s/ct/rls/>.

Sacred texts

The Noble Qur'an, University of South California, Muslim Student Association, Compendium of Muslim Texts, <http://www.usc.edu/dept/MSA/quran/>.

University of South California, Muslim Student Association, Hadith Database, including the collections *Sahih Bukhari, Sahih Muslim, Sunan Abu-Dawud* and *Malik's Muwatta*, <http://www.usc.edu/dept/MSA/reference/searchhadith.html>.

Official documents and covenants

Baker, J. A. III and Hamilton, L. H. (co-chairs), *The Iraq Study Group Report* (Iraq Study Group: 2006).

Borisov, T., '17 osobo opasnykh: publikuem spisok organizatsii, priznannykh Verkhovnym sudom Rossii terroristicheskimi' [17 most dangerous: groups listed as terrorist organizations by the Russian Supreme Court], *Rossiiskaya gazeta*, 28 July 2006.

British Home Office, 'Proscribed terrorist groups', <http://security.home office.gov.uk/legislation/current-legislation/terrorism-act-2000/proscribed-terrorist-groups>.

British House of Commons, *Report of the Official Account of the Bombings in London on 7th July 2005* (The Stationery Office: London, May 2006).

British Intelligence and Security Committee, *Report into the London Terrorist Attacks on 7 July 2005*, Cm 6785 (The Stationery Office: London, May 2006).

Covenant of the Islamic Resistance Movement [Hamas], 18 Aug. 1988, English translation available at <http://www.yale.edu/lawweb/avalon/mideast/hamas.htm>.

Europol, *EU Terrorism Situation and Trend Report 2007* (Europol: The Hague, 2007).

Sageman, M., 'Radicalization of global Islamist terrorists', Testimony before the US Senate Committee on Homeland Security and Governmental Affairs, 27 June 2007, <http://hsgac.senate.gov/index.cfm?Fuse action=Hearings.Detail&HearingID=460>.

US Department of the Army, Headquarters, *Counterinsurgency*, Field Manual no. 3-24/Marine Corps Warfighting Publication no. 3-33.5 (Department of the Army: Washington, DC, Dec. 2006).

—, *Low-Intensity Conflict*, Field Manual no. 100 20 (Government Printing Office: Washington, DC, 1981).

—, *Operational Terms and Symbols*, Field Manual no. 1 02/Marine Corps Reference Publication no. 5-2A (Department of the Army: Washington, DC, 2002).

—, *Operations in a Low-Intensity Conflict*, Field Manual no. 7-98 (Government Printing Office: Washington, DC, 1992).

US Department of Defense, *Measuring Stability and Security in Iraq*, Report to Congress (Department of Defense: Washington, DC, Mar. 2007), <http://www.defenselink.mil/home/features/Iraq_Reports/>.

US Department of State, Office of Counterterrorism, 'Foreign terrorist organizations (FTOs)', Fact sheet, Washington, DC, 11 Oct. 2005, <http://www.state.gov/s/ct/rls/fs/37191.htm>.

Fatwas, theoretical writings, reports and pamphlets

al-Ali, H. bin A., [Covenant of the Supreme Council of Jihad Groups], 13 Jan. 2007, <http://www.h-alali.net/m_open.php?id=991da3ae-f492-1029-a701-0010dc91cf69> (in Arabic).

Azzam, A., *Defence of the Muslim Lands: The First Obligation after Iman*, English translation of Arabic text (Religioscope: Fribourg, Feb. 2002), <http://www.religioscope.com/info/doc/jihad/azzam_defence_1_table.htm>.

Beam, L., 'Leaderless resistance', *The Seditionist*, no. 12 (Feb. 1992), <http://www.louisbeam.com/leaderless.htm>.

Feldner, Y., 'Debating the religious, political and moral legitimacy of suicide bombings, part 1: the debate over religious legitimacy', Inquiry and Analysis Series no. 53, Middle East Media Research Institute (MEMRI), 2 May 2001, <http://memri.org/bin/articles.cgi?Page=archives&Area=ia&ID=IA5301>.

Guevara, E. C., *Guerrilla Warfare* (Penguin: London, 1969).

Human Rights Watch, 'United States—"We are not the enemy": hate crimes against Arabs, Muslims, and those perceived to be Arab or Muslim after September 11', vol. 14, no. 6 (G) (Nov. 2002), <http://www.hrw.org/reports/2002/usahate/>.

Laden, O. bin, 'Full text: bin Laden's "letter to America"', *The Observer*, 24 Nov. 2002.

—, [World Islamic Front for jihad against Jews and crusaders: initial 'fatwa' statement], *al-Quds al-Arabi*, 23 Feb. 1998, p. 3, available at <http://www.library.cornell.edu/colldev/mideast/fatw2.htm> and in English translation at <http://www.pbs.org/newshour/terrorism/international/fatwa_1998.html>.

Laqueur, W. (ed.), *Voices of Terror: Manifestos, Writing and Manuals of Al-Qaeda, Hamas, and Other Terrorists from around the World and throughout the Ages* (Reed Press: New York, 2004).

Lawrence, B. (ed.), *Messages to the World: The Statements of Osama bin Laden* (Verso: London, 2005).

Mao Tse-tung, *On Guerrilla Warfare* (University of Illinois Press: Champaign, Ill., 2000).

Marighella, C., *Minimanual of the Urban Guerrilla* (Paladin Press: Boulder, Colo., 1975); the text is also available at <http://www.marxists.org/archive/marighella-carlos/1969/06/minimanual-urban-guerrilla/>.

Maududi, S. A. A., 'The political theory of Islam', eds Moaddel and Talattof, pp. 263–71.

—, 'Self-destructiveness of Western civilization', eds Moaddel and Talattof, pp. 325–32.

Middle East Media Research Institute (MEMRI), 'Sheikh Al-Qaradhawi on Hamas Jerusalem Day online', Special Dispatch Series no. 1051, MEMRI, 18 Dec. 2005, <http://memri.org/bin/articles.cgi?Page=archives&Area=sd&ID=SP105105>.

Moaddel, M. and Talattof, K. (eds), *Contemporary Debates in Islam: An Anthology of Modernist and Fundamentalist Thought* (Macmillan: Basingstoke, 2000)

Mujahideen Shura Council in Iraq, 'The announcement of the establishment of the Islamic State of Iraq', 15 Oct. 2006.

Muslim Public Affairs Council, 'Counterproductive counterterrorism: how anti-Islamic rhetoric is impeding America's homeland security', Dec. 2004, <http://www.mpac.org/article.php?id=354>.

Project for the Research of Islamist Movements (PRISM), New Islamist Rulings on Jihad and Terrorism, <http://www.e-prism.org/>.

Qutb, S., *Milestones* (Unity Publishing Co.: Cedar Rapids, Iowa, 1980).

—, 'War, peace, and Islamic Jihad', eds Moaddel and Talattof, pp. 223–45.

Taymiyyah, A. ibn, 'The religious and moral doctrine of jihad', reproduced in ed. Laqueur, pp. 391–93; for the complete text in English see <http://www.islamistwatch.org/texts/taymiyyah/moral/moral.html>.

'Zarqawi's pledge of allegiance to al-Qaeda: from *Mu'asker al-Battar*, issue 21', transl. J. Pool, *Terrorism Monitor*, vol. 2, no. 24 (16 Dec. 2004), pp. 4–6.

II. Literature

General

Bjørgo, T. (ed.), *Root Causes of Terrorism: Myths, Reality and Ways Forward* (Routledge: Abingdon, 2005)

Budnitsky, O. V., *Terrorizm v rossiiskom osvoboditel'nom dvizhenii: ideologiya, etika, psikhologiya (vtoraya polovina XIX–nachalo XX v.)* [Terrorism in the Russian liberation movement: ideology, ethics, psychology (the first half of the 19th–early 20th century)] (ROSSPEN: Moscow, 2000).

Byman, D., *Deadly Connections: States that Sponsor Terrorism* (Cambridge University Press: Cambridge, 2005).

Clodfelter, M., *Warfare and Armed Conflict: A Statistical Reference to Casualty and Other Figures, 1618–1991* (McFarland: Jefferson, N.C., 1992).

Crenshaw, M., 'The causes of terrorism', *Comparative Politics*, vol. 13, no. 4 (July 1981), pp. 379–99.

Eck, K. and Hultman, L., 'One-sided violence against civilians in war: insights from new fatality data', *Journal of Peace Research*, vol. 44, no. 2 (Mar. 2007), pp. 233–46.

Fedorov, A. V. (ed.), *Superterrorizm: novyi vyzov novogo veka* [Superterrorism: a new challenge of the new century] (Prava Cheloveka: Moscow, 2002).

Freedman, L. (ed.), *Superterrorism: Policy Responses* (Blackwell: Oxford, 2002).

Gurr, T. R., *Why Men Rebel* (Princeton University Press: Princeton, N.J., 1971).

Harbom, L. and Wallensteen, P., 'Armed conflict, 1989–2006', *Journal of Peace Research*, vol. 44, no. 5 (Sep. 2007), pp. 623–34.

Hardin, R., *One for All: The Logic of Group Conflict* (Princeton University Press: Princeton, N.J., 1995).

Herman, E. S. and O'Sullivan, G., '"Terrorism" as ideology and cultural industry', ed. A. George, *Western State Terrorism* (Routledge: New York, 1991), pp. 39–75.

Hoffman, B., *Inside Terrorism*, revised edn (Columbia University Press: New York, 2006).

Horne, A., *A Savage War of Peace: Algeria 1954–1962* (Macmillan: London, 1977).

Laqueur, W., *A History of Terrorism* (Transaction: New Brunswick, N.J., 2001).

Lia, B. and Skjølberg, K., *Causes of Terrorism: An Expanded and Updated Review of the Literature* (Norwegian Defence Research Establishment: Kjeller, 2005), <http://rapporter.ffi.no/rapporter/2004/04307.pdf>.

Murphy, J. F., *State Support of International Terrorism: Legal, Political, and Economic Dimensions* (Westview: Boulder, Colo., 1989).

Reno, W., *Warlord Politics and African States* (Lynne Rienner: Boulder, Colo., 1998).

Schmid, A. P. and Jongman, A. J., *Political Terrorism: A New Guide to Actors, Authors, Concepts, Data Bases, Theories, and Literature* (North-Holland: Amsterdam, 1988).

Soares, J., 'Terrorism as ideology in international relations', *Peace Review*, vol. 19, no. 1 (Jan. 2007), pp. 113–18.

Stepanova, E., *Anti-terrorism and Peace-building During and After Conflict*, SIPRI Policy Paper no. 2 (SIPRI: Stockholm, June 2003), <http://books.sipri.org/product_info?c_product_id=187>.

—, 'Trends in armed conflicts', *SIPRI Yearbook 2008: Armaments, Disarmament and International Security* (Oxford University Press: Oxford, forthcoming 2008).

Stohl, M. and Lopez G. A., *The State as Terrorist: The Dynamics of Governmental Violence and Repression* (Greenwood Press: Westport, Conn., 1984).

Sztompka, P., *The Sociology of Social Change* (Blackwell: Oxford, 1993).

University of British Columbia, Human Security Centre, *Human Security Brief 2006* (Human Security Centre: Vancouver, 2006), <http://www.humansecuritybrief.info/>.

—, *Human Security Report 2005: War and Peace in the 21st Century* (Oxford University Press: New York, 2005), <http://www.humansecurityreport.info/>.

Walker, I. and Smith H. J., *Relative Deprivation: Specification, Development, and Integration* (Cambridge University Press: Cambridge, 2001).

Terrorism, conflict and asymmetry

Aggestam, K., 'Mediating asymmetrical conflict', *Mediterranean Politics*, vol. 7, no. 1 (spring 2002), pp. 69–91.

Arreguín-Toft, I., *How the Weak Win Wars: A Theory of Asymmetric Conflict*, Cambridge Studies in International Relations no. 99 (Cambridge University Press: Cambridge, 2005).

Metz, S. and Johnson, D. V., *Asymmetry and U.S. Military Strategy: Definition, Background, and Strategic Concepts* (US Army War College, Strategic Studies Institute: Carlisle, Pa., Jan. 2001).

O'Connor, T., 'International terrorism as asymmetric warfare', 16 Dec. 2006, <http://www.apsu.edu/oconnort/3420/3420lect02.htm>.

Reynolds, J. W., *Deterring and Responding to Asymmetrical Threats* (US Army Command and General Staff College, School of Advanced Military Studies: Fort Leavenworth, Kans., 2003).

Stepanova, E., 'Terrorism as a tactic of spoilers in peace processes', eds E. Newmann and O. Richards, *Challenges to Peacebuilding: Managing Spoilers during Conflict Resolution* (United Nations University Press: Tokyo, 2006), pp. 78–104.

—, 'Terrorizm i asimmetrichnyi konflikt: problemy opredeleniya i tipologiya' [Terrorism and asymmetric conflict: problems of definition and typology], *Sovremennyi terrorizm: istoki, tendentsii, problemy preodoleniya* [Modern terrorism: sources, trends and the problems of countering], Notes of the International University in Moscow no. 6 (International University Press: Moscow, 2006), pp. 177–90.

Waldmann, P., *Terrorismus und Bürgerkrieg: der Staat in Bedrängnis* [Terrorism and civil war: the state in distress] (Gerling Akademie Verlag: Munich, 2003).

Terrorism and radical nationalism

Alonso, R., *The IRA and Armed Struggle* (Routledge: London, 2006).

Anderson, B., *Imagined Communities: Reflections on the Origin and Spread of Nationalism* (Verso: London, 1991).

Brubaker, R., *Ethnicity Without Groups* (Harvard University Press: Cambridge, Mass., 2004).

— and Laitin, D. D., 'Ethnic and nationalist violence', *Annual Review of Sociology*, vol. 24 (1998), pp. 423–52

Byman, D., 'The logic of ethnic terrorism', *Studies in Conflict and Terrorism*, vol. 21, no. 2 (Apr.–June 1998), pp. 149–70.

Chirot, D. and Seligman, M. E. P. (eds), *Ethnopolitical Warfare: Causes, Consequences, and Possible Solutions* (American Psychological Association: Washington, DC, 2000).

Coakley, J., *The Territorial Management of Ethnic Conflict*, 2nd edn (Frank Cass: London, 2003).

Connor, W., *Ethnonationalism: The Quest for Understanding* (Princeton University Press: Princeton, N.J., 1994).

De Vries, H. and Weber, S. (eds), *Violence, Identity, and Self-Determination* (Stanford University Press: Palo Alto, Calif., 1997).

Fearon, J. D. and Laitin, D. D., 'Ethnicity, insurgency and civil war', *American Political Science Review*, vol. 97, no. 1 (Feb. 2003), pp. 75–90.

— and —, 'Explaining interethnic cooperation', *American Political Science Review*, vol. 90, no. 4 (Dec. 1996), pp. 715–35.

Galula, D., *Pacification in Algeria, 1956–1958*, new edn (RAND: Santa Monica, Calif., 2006).

Gellner, E., *Nations and Nationalism* (Blackwell: Oxford, 1981).

Gurr, T. R., 'Ethnic warfare on the wane', *Foreign Affairs*, vol. 79, no. 3 (May/June 2000), pp. 52–64.

Hobsbaum, E., *Nations and Nationalism Since 1780: Programme, Myth, Reality* (Cambridge University Press: Cambridge, 1990).

— and Ranger, T. (eds), *The Invention of Tradition* (Cambridge University Press: Cambridge, 1983).

Horowitz, D. L., *Ethnic Groups in Conflict* (University of California Press: Berkeley, Calif., 1985).

Ignatieff, M., *Blood and Belonging: Journeys into the New Nationalism* (Farrar, Straus and Giroux: New York, 1993).

Irvin, C. L., *Militant Nationalism: Between Movement and Party in Ireland and the Basque Country* (University of Minnesota Press: Minneapolis, Minn.,1999).

Kaplan, R. D., *The Ends of the Earth: A Journey to the Frontiers of Anarchy* (Random House: New York, 1996).

Lefebvre, S., *Perspectives on Ethno-nationalist/Separatist Terrorism* (Defence Academy of the United Kingdom, Conflict Studies Research Centre: Camberley, May 2003).

McGarry, J. and O'Leary, B., 'Eliminating and managing ethnic differences', eds J. Hutchinson and A. D. Smith, *Ethnicity* (Oxford University Press: Oxford, 1996), pp. 333–40.

— and — (eds), *The Politics of Ethnic Conflict Regulation: Case Studies of Protracted Ethnic Conflicts* (Routledge: London, 1993).

Moore, M. (ed.), *National Self-Determination and Secession* (Oxford University Press: Oxford, 1998).

Motyl, A. J. (ed.), *Encyclopedia of Nationalism* (Academic Press: San Diego, Calif., 2000).

Mueller, J., 'The banality of "ethnic war"', *International Security*, vol. 25, no. 1 (summer 2000), pp. 42–70.

Reinares, F., *Patriotas de la Muerte: Quiénes han militado en ETA y por qué* [Patriots of death: who joined ETA and why] (Taurus: Madrid, 2001).

Sambanis, N., 'Do ethnic and nonethnic civil wars have the same causes? A theoretical and empirical inquiry (part 1)', *Journal of Conflict Resolution*, vol. 45, no. 3 (June 2001), pp. 259–82.

Schaeffer, R. K., *Severed States: Dilemmas of Democracy in a Divided World* (Rowman & Littlefield: Lanham, Md., 1999).

Simpson, G. J., 'The diffusion of sovereignty: self-determination in the post-colonial age', *Stanford Journal of International Law*, vol. 32 (1996), pp. 255–86.

Smith, A.D., *The Ethnic Origins of Nations* (Blackwell: Oxford, 1988).

—, *Nationalism: Theory, Ideology, History* (Polity: Cambridge, 2001).

Strmiska, M., 'Political radicalism, subversion and terrorist violence in democratic systems', *Středoevropské politické studie/Central European Political Studies Review*, vol. 2, no. 3 (summer 2000), pp. 50–59.

Tilly, C., 'National self-determination as a problem for all of us', *Daedalus*, vol. 122, no. 3 (summer 1993), pp. 29–36.

—, *The Politics of Collective Violence* (Cambridge University Press: Cambridge, 2003).

Tishkov, V., *Chechnya: Life in a War-Torn Society* (University of California Press: Berkeley, Calif., 2004).

Volkan, V., *Bloodlines: From Ethnic Pride to Ethnic Terrorism* (Westview Press: Boulder, Colo., 1997).

Waldmann, P., *Ethnischer Radikalismus: Ursachen und Folgen gewaltsamer Minderheitenkonflikte am Beispiel des Baskenlandes, Nordirlands und Quebecs* [Ethnic radicalism: causes and consequences of violent minority conflicts through the examples of the Basque country, Northern Ireland and Quebec] (Westdeutscher Verlag: Opladen, 1992).

Terrorism and religious and quasi-religious extremism

Ayoob, M. (ed.), *The Politics of Islamic Reassertion* (St. Martin's Press: New York, 1981).

Ayubi, N. N., *Political Islam: Religion and Politics in the Arab World* (Routledge: London, 1991).

Barton, G., *Jemaah Islamiyah: Radical Islam in Indonesia* (Singapore University Press: Singapore, 2005).

Benjamin, D. and Simon, S., *The Age of Sacred Terror* (Random House: New York, 2002).

Blanchard, C. M., US Congress, Congressional Research Service (CRS), *Al-Qaeda: Statements and Evolving Ideology*, CRS Report for Congress RL32759 (CRS: Washington, DC, 9 July 2007).

Bokhari, L. et al., *Paths to Global Jihad: Radicalisation and Recruitment to Terror Networks*, Norwegian Defence Research Establishment (FFI) Report no. 2006/00935 (Norwegian Defence Research Establishment: Kjeller, 2006).

Burgat, F., *Face to Face with Political Islam* (I. B.Taurus: London, 1997).

Council on Foreign Relations, 'Hamas', Backgrounder, 8 June 2007, <http://www.cfr.org/publication/8968>.

Esposito, J. L. (ed.), *The Oxford Dictionary of Islam* (Oxford University Press: Oxford, 2003).

— (ed.), *Political Islam: Revolution, Radicalism or Reform* (Lynne Rienner: Boulder, Colo., 1997).

—, *Unholy War: Terror in the Name of Islam* (Oxford University Press: Oxford, 2002).

Hall, J., Schuyler, P. D. and Trin, S., *Apocalypse Observed: Religious Movements and Violence in North America, Europe, and Japan* (Routledge: London, 2000).

Hamzeh, A. N., 'Islamism in Lebanon: a guide to the groups', *Middle East Quarterly*, vol. 4, no. 3 (Sep. 1997), pp. 47–54.

—, 'Lebanon's Hizbullah: from Islamic revolution to parliamentary accommodation', *Third World Quarterly*, vol. 14, no. 2 (Apr. 1993), pp. 321–37.

Hoffman, B., '"Holy terror": the implications of terrorism motivated by a religious imperative', *Studies in Conflict and Terrorism*, vol. 18, no. 4 (Oct.–Dec. 1995), pp. 271–84.

—, 'Old madness, new methods: revival of religious terrorism begs for broader U.S. policy', *RAND Review*, vol. 22, no. 2 (winter 1998/99).

International Crisis Group (ICG), *Islamism, Violence and Reform in Algeria: Turning the Page*, Middle East Report no. 29 (ICG: Brussels, 30 July 2004), <http://www.crisisgroup.org/home/index.cfm?id=2884>.

—, *Recycling Militants in Indonesia: Darul Islam and the Australian Embassy Bombing*, ICG Asia Report no. 92 (ICG: Brussels, 22 Feb. 2005), <http://www.crisisgroup.org/home/index.cfm?id=3280>.

—, *Understanding Islamism*, Middle East/North Africa Report no. 37 (ICG: Brussels, 2 Mar. 2005), <http://www.crisisgroup.org/home/index.cfm?id=3301>.

Juergensmeyer, M., *Terror in the Mind of God: The Global Rise of Religious Violence* (University of California Press: Berkeley, Calif., 2000).

Kepel, G., *Jihad: The Trail of Political Islam* (I. B. Tauris: London, 2004).

—, *The Prophet and the Pharaoh: Muslim Extremism in Egypt* (Saqi: London, 1985).

—, *The Roots of Radical Islam* (Saqi: London, 2005).

Lakhdar, L., 'The role of fatwas in incitement to terrorism', Special Dispatch Series no. 333, Middle East Media Research Institute (MEMRI), 18 Jan. 2002, <http://memri.org/bin/articles.cgi?Page=archives&Area=sd&ID=SP33302>.

Malashenko, A., 'Brodit li prizrak "islamskoi ugrozy"?' [Is the spectre of 'Islamic threat' roaming?], Carnegie Moscow Center Working Paper no. 2/2004, Moscow, 2004, <http://www.carnegie.ru/ru/pubs/workpapers/70269.htm>.

—, *Islamskie orientiry Severnogo Kavkaza* [Islamic factor in the North Caucasus] (Carnegie Moscow Center/Gendalf: Moscow, 2001).

—, *Islamskoe vozrozhdenie v sovremennoi Rossii* [The Islamic renaissance in contemporary Russia] (Carnegie Moscow Center: Moscow, 1998).

Mishal, S. and Sela, A., *The Palestinian Hamas: Vision, Violence, and Coexistence* (Columbia University Press: New York, 2000).

Moussalli, A. S., *Radical Islamic Fundamentalism: The Ideological and Political Discourse of Sayyid Qutb* (American University of Beirut: Beirut, 1992).

Naumkin, V. V., *Militant Islam in Central Asia: The Case of the Islamic Movement of Uzbekistan* (University of California, Berkeley, Berkeley Institute of Slavic, East European and Eurasian Studies: Berkeley, Calif., 2003), <http://repositories.cdlib.org/iseees/bps/2003_06-naum/>.

Paz, R., 'Catch as much as you can: Hasan al-Qaed (Abu Yahya al-Libi) on Jihadi terrorism against Muslims in Muslim countries', PRISM Occasional Papers, vol. 5, no. 2 (Aug. 2007), <http://www.e-prism.org/projects andproducts.html>.

—, 'Islamic legitimacy for the London bombings', PRISM Occasional Papers, vol. 3, no. 4 (July 2005), <http://www.e-prism.org/projectsand products.html>.

Ranstorp, M., *Hizb'Allah in Lebanon: The Politics of the Western Hostage Crisis* (St Martin's Press: New York, 1997).

—, 'Terrorism in the name of religion', *Journal of International Affairs*, vol. 50, no. 1 (summer 1996), pp. 41–62.

Rapoport, D. C., 'Fear and trembling: terrorism in three religious traditions', *American Political Science Review*, vol. 78, no. 3 (Sep. 1984), pp. 658–77.

'Religion and terrorism: interview with Dr. Bruce Hoffman', Religioscope, 22 Feb. 2002, <http://www.religioscope.com/info/articles/003_Hoffman_ terrorism.htm>.

Roy, O., *The Failure of Political Islam* (Harvard University Press: Cambridge, Mass.,1994).

Rubin, B. (ed.), *Revolutionaries and Reformers: Contemporary Islamist Movements in the Middle East* (State University of New York Press: Albany, N.Y., 2003).

Saad-Ghorayeb, A., *Hizbu'llah: Politics and Religion* (Pluto Press: London, 2002).

Tibi, B., *The Challenge of Fundamentalism: Political Islam and the New World Disorder* (University of California Press: Berkeley, Calif., 1998).

Wiktorowicz, Q. (ed.), *Islamic Activism: A Social Movement Theory Approach* (Indiana University Press: Bloomington, Ind., 2004).

Structures of terrorist groups

Arquilla, J. and Karasik, T., 'Chechnya: a glimpse of future conflict?', *Studies in Conflict and Terrorism*, vol. 22, no. 3 (July–Sep. 1999), pp. 207–29.

— and Ronfeldt, D., 'Netwar revisited: the fight for the future continues', *Low Intensity Conflict & Law Enforcement*, vol. 11, nos 2–3 (winter 2002), pp. 178–89.

— and — (eds), *Networks and Netwars: The Future of Terror, Crime, and Militancy* (RAND: Santa Monica, Calif., 2001).

— and —, *Swarming and The Future of Conflict*, RAND Documented Briefing (RAND: Santa Monica, Calif., 2000).

Baddeley, J. F., *The Russian Conquest of the Caucasus* (Longmans, Green, and Co.: London, 1908).

Castells, M., *The Information Age: Economy, Society and Culture*, vol. 1, *The Rise of the Network Society*, and vol. 2, *The Power of Identity*, 2nd edn (Blackwell: Oxford, 2000).

Galeotti, M., '"Brotherhoods" and "associates": Chechen networks of crime and resistance', *Low Intensity Conflict & Law Enforcement*, vol. 11, no. 2/3 (winter 2002), pp. 340–52.

Gammer, M., *The Lone Wolf and the Bear: Three Centuries of Chechen Defiance of Russian Rule* (University of Pittsburgh Press: Pittsburgh, Pa., 2006).

Garfinkel, S. L., 'Leaderless resistance today', *First Monday*, vol. 8, no. 3 (Mar. 2003), <http://firstmonday.org/issues/issue8_3/>.

Gerlach, L., 'Protest movements and the construction of risk', eds B. B. Johnson and V. T. Covello, *The Social and Cultural Construction of Risk: Essays on Risk Selection and Perception* (D. Reidel: Boston, Mass., 1987), pp. 103–45.

— and Hine, V., *People, Power, Change: Movements of Social Transformation* (Bobbs-Merril Co.: New York, 1970).

Gunaratna, R., *Inside Al Qaeda: Global Network of Terror* (Columbia University Press: New York, 2002).

International Crisis Group (ICG), *Indonesia Backgrounder: How the* Jemaah Islamiyah *Terrorist Network Operates*, Asia Report no. 43 (ICG: Brussels, 11 Dec. 2002), <http://www.crisisgroup.org/home/index.cfm?id=1397>.

Kaplan, J., 'Leaderless resistance', *Terrorism and Political Violence*, vol. 9, no. 3 (autumn 1997), pp. 80–95.

Kulikov, S. A. and Love, R. R., 'Insurgent groups in Chechnya', *Military Review*, vol. 83, no. 6 (Nov./Dec. 2003), pp. 21–29.

Lesser, I. O. et al., *Countering the New Terrorism* (RAND: Santa Monica, Calif., 1999).

McCalister, W. S., 'The Iraq insurgency: anatomy of a tribal rebellion', *First Monday*, vol. 10, no. 3 (Mar. 2005), <http://firstmonday.org/issues/issue 10_3/>.

Mayntz, R., *Organizational Forms of Terrorism: Hierarchy, Network, or a Type Sui Generis?*, Max Planck Institute for the Study of Societies (MPIfG) Discussion Paper no. 04/4 (MPIfG: Cologne, 2004), <http://edoc.mpg.de/230590>.

Nohria, N. and Eccles, R. G. (eds), *Networks and Organizations: Structure, Form, and Action* (Harvard Business School Press: Boston, Mass., 1992).

O'Brien, B., *Long War: IRA and Sinn Fein 1985 to Today* (Syracuse University Press: Syracuse, N.Y., 1999).

Ouchi, W. G., 'Markets, bureaucracies and clans', *Administrative Science Quarterly*, vol. 25, no. 1 (Mar. 1980), pp. 129–41.

Paz, R., 'A global jihadi umbrella for strategy and ideology: the Covenant of the Supreme Council of Jihad Groups', PRISM Occasional Papers, vol. 5, no. 1 (Jan. 2007), <http://www.e-prism.org/projectsandproducts.html>.

Ronfeldt, D., 'Al Qaeda and its affiliates: a global tribe waging segmental warfare?', *First Monday*, vol. 10, no. 3 (Mar. 2005), <http://firstmonday.org/issues/issue10_3/>.

Sageman, M., *Leaderless Jihad: Terror Networks in the Twenty-First Century* (University of Pennsylvania Press: Philadelphia, Pa., 2007).

—, 'Understanding terror networks', Foreign Policy Research Institute E-Notes, 1 Nov. 2004, <http://www.fpri.org/enotes/pastenotes.html>.

—, *Understanding Terror Networks* (University of Pennsylvania Press: Philadelphia, Pa., 2004).

Scott, J., *Social Network Analysis: A Handbook*, 2nd edn (Sage: London, 2000).

Stepanova, E., 'Organizatsionnyie formy global'nogo dzhikhada' [Organizational forms of global jihad], *Mezhdunarodnye protsessy*, vol. 4, no. 1 (10) (Jan.-Apr. 2006), <http://www.intertrends.ru/tenth.htm>.

Taarnby, M., 'Understanding recruitment of Islamist terrorists in Europe', ed. M. Ranstorp, *Mapping Terrorism Research: State of the Art, Gaps and Future Direction* (Routledge: London, 2007), pp. 164–86.

Thomas, T. L., 'The battle of Grozny: deadly classroom for urban combat', *Parameters*, vol. 29, no. 2 (summer 1999), pp. 87–102.

Tsoukas, H. and Knudsen, C. (eds), *The Oxford Handbook of Organization Theory* (Oxford University Press: Oxford, 2005).

Weber, M., *The Theory of Social and Economic Organization*, transl. A. M. Henderson and T. Parsons (Free Press: Glencoe, Ill., 1947).

Index